COMPREHENDING AND DECODING THE COSMOS

Discovering Solutions To Over A Dozen Cosmic Mysteries By Utilizing Dark Matter Relationism, Cosmology, And Astrophysics

JEROME DREXLER

Universal Publishers
USA • 2006

COMPREHENDING AND DECODING THE COSMOS:
Discovering Solutions To Over A Dozen Cosmic Mysteries By Utilizing Dark Matter Relationism, Cosmology, And Astrophysics

Copyright © 2006 Jerome Drexler
All rights reserved.

This book is protected by copyright. No part of it may be reproduced or translated without the prior written permission of the copyright owner, except as permitted by law.

Universal Publishers
Boca Raton, Florida • USA
2006

ISBN: 1-58112-929-7

www.universal-publishers.com

This book is dedicated to Sylvia, my wife, friend, and lifelong partner.

"Leave the beaten track occasionally and dive into the woods. Every time you do so you will find something you have never seen before. Follow it up, explore all around it, and before you know it, you will have something to think about to occupy your mind. All really big discoveries are the result of thought."

~ Alexander Graham Bell ~

"It is dangerous to be right in matters on which the established authorities are wrong."

~ Voltaire ~

PREFACE

One hundred years ago, Albert Einstein announced the Special Theory of Relativity, which predicted and explained that a proton traveling near the speed of light could have a relativistic mass a thousand, a million, or even a billion times greater than the mass of a proton at rest. Therefore, the gravitational strength of the multitudinous galaxy-orbiting relativistic protons moving in the cosmos could create extremely large gravity-related tidal forces on nearby matter.

Ever since astronomer Fritz Zwicky discovered the presence of dark matter (DM) in the Coma cluster of galaxies in 1933 and astronomer Vera Rubin confirmed the existence of dark matter halos around galaxies in 1977, cosmologists and astrophysicists have been trying to identify the dark matter particles.

In 1984, scientists developed a Cold Dark Matter (CDM) theory based upon a theoretical uncharged, slow moving particle, they called the Weakly Interacting Massive Particle (WIMP). More recently, it was determined that the theoretical WIMP dark matter particles would require a mass of 35 to 10,000 times greater than the mass of a proton at rest in order to exhibit the observed gravity-related forces of dark matter halos. However, searches for the theoretical WIMP particles during the past 20 years have all come up empty handed.

For this reason, and knowing that the relativistic proton easily could meet the mass requirement of the mysterious dark matter particles and that relativistic cosmic ray protons are widely observed, I have endeavored to bring the relativistic proton dark matter theory/cosmology to the attention of dark matter astronomers, astrophysicists, and cosmologists as well as to NASA, the National Science Foundation (NSF), and the U.S. Department of Energy (DOE) through two recent publications. My earlier astrophysics/cosmology book, *"How Dark Matter Created Dark Energy And The Sun,"* was published in December 2003; and on April 22, 2005, my 19-page follow-up paper was posted on the Cornell University Library's arXiv.gov website as e-Print No. astro-ph/0504512. This paper is entitled, "Identifying Dark Matter Through The Constraints Imposed by Fourteen Astronomically Based 'Cosmic

Constituents'" and is available at http://arxiv.org/ftp/astro-ph/papers/0504/0504512.pdf.

The principal objective of this book is found in its main title, *"Comprehending And Decoding The Cosmos."* The subtitle describes the concepts and goals of the book, namely, *"Discovering Solutions To Over A Dozen Cosmic Mysteries By Utilizing Dark Matter Relationism, Cosmology, And Astrophysics."* *Dark matter relationism* is epitomized in my use of cosmic mysteries and *relationism* to tentatively identify dark matter and then to confirm its validity by using this same dark matter candidate to provide plausible explanations for additional cosmic mysteries, including some previously *not known to be related to dark matter*.

On June 22, 2003, Jerimiah P. Ostriker of the Department of Astrophysical Sciences and Paul J. Steinhardt of the Department of Physics at Princeton University published a paper in the journal Science entitled, "New Light on Dark Matter." In a sense, this book responds to the final paragraph of their conclusion section, which reads:

> We have sketched out the kinds of astronomical tests that could be done to narrow the search [for dark matter],

but if history teaches us anything it is that the next important clues will come from a surprising direction. Some observation or calculation will be made that will reorient our inquiries and, if this happens as has happened so often in the past, we will realize that the important evidence has been sitting unnoticed under our noses for decades.

This book, *"Comprehending And Decoding The Cosmos,"* deviates significantly from mainstream cosmological and astrophysical theories. My seven years as a Member of the Technical staff at Bell Laboratories taught me to think outside the box and to become a prolific inventor, utilizing applied physics. Alexander Graham Bell's philosophy on invention, displayed on an engraved bronze plaque at Bell Labs, left a lasting impression on me:

Leave the beaten track occasionally and dive into the woods. Every time you do so you will find something you have never seen before.
Follow it up, explore all around it, and before you know it, you will have something to think about to occupy your mind.
All really big discoveries are the result of thought.

CONTENTS

INTRODUCTION   17

PART I

CHAPTER 1: The Search For The Identity Of Dark Matter (DM)   27

CHAPTER 2: Additional Approaches To DM Research   35

CHAPTER 3: DM Research Guided By The Three Related Hypotheses   37

CHAPTER 4: SigChar A - DM Proton Energies   43

CHAPTER 5: SigChar B - The Milky Way's Magnetic Fields   45

CHAPTER 6: SigChar C - Larmor Radius Equation   47

CHAPTER 7: SigChar D - The Milky Way's DM Halos And Proton Energies   49

CHAPTER 8: SigChar E - Paths Of Protons   51

CHAPTER 9: SigChar F - Proton Streams Creating Magnetic Fields   53

CHAPTER 10: SigChar G - Proton Flux And Kinetic Energy In Halos   55

CHAPTER 11: SigChar H - Proton Relativistic Mass Losses From Synchrotron Radiation   57

CHAPTER 12: SigChar I - Magnetic Bulges Leading To Increased Synchrotron Radiation From Protons ... 59

CHAPTER 13: SigChar J - Why DM Halo Protons Enter Their Enclosed Galaxy And Lose Relativistic Mass ... 61

CHAPTER 14: SigChar K - Protons/Helium Nuclei Collisions With Hydrogen Clouds ... 63

CHAPTER 15: SigChar L - Linearly Rising Rotation Curves Indicating That Low-Surface Brightness (LSB) Dwarf Galaxy DM Halos Are "Weakly Centrally Concentrated" (That Is, "Hollow") ... 65

CHAPTER 16: SigChar M - Explanation For The Two "Knees" And "Ankle" Of The Cosmic Ray Energy Distribution ... 69

CHAPTER 17: SigChar N - Proton Synchrotron Radiation Losses And Proton Collision Losses Possibly Could Lead To An Accelerating Expansion Of The Universe ... 71

CHAPTER 18: SigChar O - Radiating DM Halo Protons Become Cosmic Ray Protons ... 75

CHAPTER 19: SigChar P - Long, Large DM Filaments Creating Galaxy Clusters ... 77

CHAPTER 20: SigChar Q - Mature Galaxies In A Young Universe ... 79

CHAPTER 21: SigChar R - Conservation Of Angular Momentum ... 81

CHAPTER 22: SigChar S - No DM Cusps In The Nuclei Of Spiral Galaxies ... 83

CHAPTER 23: SigChar T - Explanations For LSB Dwarf Galaxies' Low Star Formation Rates (SFRs) And For Massive Galaxies' Very High SFRs   85

CHAPTER 24: SigChar U - The Relativistic Energy Of All The Protons In The Universe May Provide The Energy For An Accelerating Expansion Of The Universe   91

CHAPTER 25: SigChar V - Linking Relativistic DM And Dark Energy   93

CHAPTER 26: SigChar W - How The First-Generation Stars May Have Been Ignited Without Dust Or Molecular Hydrogen   95

CHAPTER 27: SigChar X - How The Later Generations Of New Stars May Have Been Ignited Utilizing Both Dust And Molecular Hydrogen   103

TABLE 1: Recap Of Signature Characteristics A - X Of Galaxy-Orbiting Relativistic Protons   106

CHAPTER 28: Tentative Conclusions, Insights, Explanations, And Possible Astrophysical Discoveries   109

PART II

CHAPTER 29: Cosmic DM Mystery #1 - Spiral Disk Galaxies Have Spherical Dark Matter Halos. Relativistic Proton DM Particles Could Form Spherical DM Halos Around Spiral Galaxies And DM Halos Around Galaxy Clusters   117

CHAPTER 30: Cosmic DM Mystery #2 - Accelerating Expansion Via Conserving DM Momentum. Relativistic Proton DM Particles Could Cause Accelerating Expansion Of The Universe And Possibly Store Dark Energy   119

CHAPTER 31: Cosmic DM Mystery #3 - Hydrogen Derived From DM Cosmic Ray Protons. Relativistic Proton DM Particles Could Be Transformed Into Low-Velocity Hydrogen, Protons, Or Proton Cosmic Rays 121

CHAPTER 32: Cosmic DM Mystery #4 - Magnetic Fields Derived From DM Cosmic Ray Protons. Relativistic Proton DM Particles Could Create The Magnetic Fields Within And Around Spiral Galaxies 123

CHAPTER 33: Cosmic DM Mystery #5 - Intersecting DM Filaments Create Galaxy Clusters. Relativistic Proton DM Particles Could Be Concentrated In The Long, Large Filaments Of DM, Which Form Galaxy Clusters Where The DM Filaments Intersect 125

CHAPTER 34: Cosmic DM Mystery #6 - Mature Galaxies Discovered In The Very Early Universe. Relativistic Proton DM Particles Could Create Large, Mature, Spiral Galaxies Less Than 2.5 Billion Years After The Big Bang 127

CHAPTER 35: Cosmic DM Mystery #7 - Dark Matter Spherical Cored Halos Have "Hollow" Cores. Relativistic Proton DM Particles Could Create Spherical DM Halos Having Predictable Outer And "Hollow" Core Diameters 129

CHAPTER 36: Cosmic DM Mystery #8 - Source Of Spiral Galaxies'/Halos' Angular Momentum. Relativistic Proton DM Particles Could Provide Angular Momentum To Spiral Galaxies And Their DM Halos 131

CHAPTER 37: Cosmic DM Mystery #9 - No Central Dark Matter Cusp Found In Spiral Galaxies. Relativistic Proton DM Particles Could Create Galaxies Without A Central DM Density Cusp 133

CHAPTER 38: Cosmic DM Mystery #10 - LSB Dwarf Galaxies Have Low Star Formation Rates. Relativistic Proton DM Particles Could Create A Starless Galaxy Or An LSB Dwarf Galaxy With Low SFRs   135

CHAPTER 39: Cosmic DM Mystery #11 - LSB Galaxies Have Inclining Star Rotation Curves. Relativistic Proton DM Particles Could Lead To Linearly Rising Rotation Curves For LSB Dwarf Galaxies And To Flat Rotation Curves For Spiral Galaxies   137

CHAPTER 40: Cosmic DM Mystery #12 - Galaxy Hydrogen Is Replenished From Halo Dark Matter. Relativistic Proton DM Particles Could Form About 80% To 85% Of The Mass Of The Universe, The Remainder Being Hydrogen, Helium, Etc.   141

CHAPTER 41: Cosmic DM Mystery #13 - Dark Matter, Hydrogen, Helium, And Muons Create Stars. Relativistic Proton DM Particles Could Ignite Hydrogen Fusion Reactions Of First-Generation Stars Using Only Hydrogen And Helium Atoms, And Of Second-Generation Stars Using Hydrogen Molecules, Helium, And Dust As Well   143

CHAPTER 42: Cosmic DM Mystery #14 - Earthbound Cosmic Ray Protons Depart From 4 Locations. Relativistic Proton DM Particles Could Create The First "Knee" At 3×10^{15} eV, The Second "Knee" Between 10^{17} eV And 10^{18} eV, And The Ankle At 3×10^{18} eV Of The Cosmic Ray Energy Distribution Near The Earth   149

TABLE 2: Recap Of Cosmic DM Mysteries #1 - #14 For Decoding The Cosmos Via DM Relationism And By Solving The Cosmic DM Mysteries   152

CHAPTER 43: Some Tentative Conclusions After The Study Of The First 14 Of The 25 Cosmic DM Mysteries - Dark Matter Relativistic Protons Appear To Be A Much Stronger DM Candidate Than The Cold Dark Matter (CDM) Uncharged WIMPs And Neutralinos, For A Number Of Reasons   153

PART III

CHAPTER 44: Cosmic DM Mystery #15 - Astrophysical Emergence Of Dark Matter Halos, After Eons. Astrophysical Emergence Of DM Halos And Long, Large, DM Filaments Could Place Constraints On The Identity Of DM Particles   157

CHAPTER 45: Cosmic DM Mystery #16 - UHECRs Arrive At Earth From Galaxy Superclusters. Ultra-High Energy Cosmic Ray Protons Arriving At Earth Probably Departed From A Galaxy Supercluster Or A Massive Galaxy Cluster   163

CHAPTER 46: Cosmic DM Mystery #17 - Starburst Galaxies Form Via Merging Galaxy Clusters. The Merging Of Spiral Galaxy Clusters Create Starburst Galaxies That Exhibit Star Formation Rates (SFRs) As Much As 50 Times Higher Than The SFR Of Spiral Galaxies   167

CHAPTER 47: Cosmic DM Mystery #18 - UHECR Protons Via Starburst Galaxies/Merging Galaxies. Spiral Galaxy Clusters, Merging To Form Starburst Galaxies, Were Recently Identified As A source Of Ultra-High Energy Cosmic Ray Protons   173

CHAPTER 48: Cosmic DM Mystery #19 - Blue Stars In Spiral Arms Vs. Red Stars In Galaxy Nucleus. The Spiral Arms Of Spiral Galaxies Contain Many Hot Blue

And Blue-White Stars Less Than One Million Years Old, And In The Galaxy Nucleus There Are Red Stars About Five Billion Years Old 177

CHAPTER 49: Cosmic DM Mystery #20 - Magnetic Field, DM Proton Energies Set Galaxy Halo Size. The Only DM Particle Candidate That "Predicts" The Size Of The Milky Way's DM Halo Is The Relativistic Cosmic Ray Proton Moving In The Extragalactic (Intergalactic) Magnetic Field Having A Strength Of About 1×10^{-9} Gauss 183

CHAPTER 50: Cosmic DM Mystery #21 - Different Dark Matter For Small Galaxies And For Clusters. Two Different Types Of DM Halo Particles Reported For Smaller Galaxies And For Galaxy Clusters 187

CHAPTER 51: Cosmic DM Mystery #22 - 800 Galaxies Detected, Less Than 1.2 Billion Years Old. Report Of Over 1,000 Clumps Of DM, With Most Harboring Several Newborn Galaxies, 12 Billion Light-Years Away 191

CHAPTER 52: Cosmic DM Mystery #23 - Fine Balance Between Dark And Baryonic Matter In Spirals. "A Fine Balance Between Dark Matter And Baryonic Matter Is Observed In Spiral Galaxies. As The Contribution Of The Baryons To The Total Rotation Velocity Increases, The Contribution Of The Dark Matter Decreases By A Compensating Amount." 195

CHAPTER 53: Cosmic DM Mystery #24 - Schmidt Law: SFR Vs. Surface Hydrogen Molecular Density. One Of The Mysteries Of Observed Isolated Spiral Galaxies Has Been The Empirical Schmidt Law Correlation Between Star Formation Rate And The Average Molecular Hydrogen Surface Density On Kiloparsec Scales 201

CHAPTER 54: Cosmic DM Mystery #25 - Mass-And-Time-Dependent SFR Graphs For Field Galaxies. The Two-Part Mystery Of Recently Observed Star-Forming Galaxies Is That Large Massive Galaxies Form Stars Early And Rapidly, But Eventually Their SFRs Fall Rapidly, Whereas Small Galaxies Form Stars Slowly Over Longer Time Scales And Their SFRs Decline Slowly Over Longer Time Scales   207

TABLE 3: Recap Of Cosmic DM Mysteries #15 - #25 For Decoding The Cosmos Via DM Relationism And By Solving The Cosmic DM Mysteries   212

CHAPTER 55: Some Conclusions And Considerations   213

CHAPTER 56: Epilogue -The Local Group's Dwarf Spheroidal Satellite Galaxies Help Define DM   221

ACKNOWLEDGMENTS   233

APPENDIX A: The Scientific Community's Long-Held Objections To Any Proton DM Theory   235

APPENDIX B: Excerpts From "*How Dark Matter Created Dark Energy And The Sun*"   241

REFERENCES   259

BIBLIOGRAPHY AND SUGGESTED SOURCES   265

GLOSSARY   269

INDEX   287

ABOUT THE AUTHOR   295

COMPREHENDING AND DECODING THE COSMOS

DISCOVERING SOLUTIONS TO OVER A DOZEN COSMIC MYSTERIES BY UTILIZING DARK MATTER RELATIONISM, COSMOLOGY, AND ASTROPHYSICS

INTRODUCTION

The author believes that the galaxy-orbiting relativistic proton appears to have the necessary characteristics of the long-sought dark matter (DM) particles, which are estimated by most scientists to comprise 80% to 85% of the total mass of the Universe. Relativistic protons do have the required mass and the required difficulty of detection and can transform themselves into hydrogen, the principal matter of galaxies, by creating and combining with electrons.

Therefore, they are capable of forming (1) galaxies and their dark matter halos, (2) galaxy clusters and their dark matter halos, (3) the long, large, filamentary dark matter that crisscrosses the cosmos, and (4) also may be capable of igniting the hydrogen fusion reaction in newborn stars.

However, for this proton-based dark matter theory to become widely accepted, there also should be astronomical evidence of multitudinous relativistic protons within the spherical dark matter halo surrounding the Milky Way (and the Earth). The author believes that the cosmic ray relativistic protons bombarding Earth every day, uniformly from all directions, go a long way toward providing such astronomical evidence.

The author estimates that the number of relativistic protons orbiting the Milky Way within the dark matter halo is roughly about 11 orders of magnitude greater than the number of cosmic ray protons plunging into all the star systems of the Milky Way each year. This estimate assumes that perhaps about 10% of the relativistic proton dark matter of the Universe was converted into ordinary hydrogen during the past 10 billion years.

Mankind has not previously explained dark matter, the accelerating expansion of the Universe, the "knees" and "ankle" of the cosmic ray proton energy distribution graph, the low star formation rates of low surface brightness (LSB) dwarf galaxies, the ignition of hydrogen fusion reactions in the first generation stars, or the departing locations of Earthbound high-energy cosmic ray protons.

A new research hypothesis has been developed by the author based upon finding astronomically based "Cosmic DM Mysteries" (Cosmic Dark-Matter-Related Mysteries) of the Universe that may be created or influenced by or have a special relationship with the dark matter halos around galaxies and galaxy clusters. Cosmic DM Mysteries have to do with mysteries or unexplained phenomena regarding celestial bodies or cosmic matter such as their shape, mass distribution, particle abundance ratios, dimensions, density, location, maturity, acceleration, velocity, linear or angular momentum, particle energies, star rotation curves, hydrogen fusion reactions, particle energy distributions, particle transformations, star ignition, and star formation rates.

Since dark matter represents 80% to 85% of the mass of the Universe, it should not be surprising that it would have an influence on or a relationship with a number of the above-indicated types of Cosmic DM Mysteries. To date, the vast majority of research conducted on dark matter by others has had to do with trying to identify the particles that comprise dark matter or to determine their gravitational effect on galaxy star rotation curves. This primarily inward-looking approach to identify the particle composition of a medium is known as *reductionism,* which is a procedure or theory that reduces or attempts to reduce complex data or phenomena to simple elements or terms.

Reductionism does not always work in physics. Many times simple entities or particles can form complex forms or combinations that have characteristics seemingly unrelated to the characteristics of the original simple entities. A hurricane is one well-known example of complex behavior whose characteristics cannot be predicted by an analysis of all the known simple entities involved in its makeup. Thus, the reductionism approach does not explain or predict the nature of a hurricane.

An alternative to reductionism is an outward-looking, cosmological-like approach that the author has developed and designated *relationism,* where a phenomenon such as the dark matter can be analyzed and categorized in terms of the Cosmic DM Mysteries that may have a relationship with dark matter. That is, in matters of the Universe, reductionism is primarily an inward-looking particle physics approach, and relationism is primarily an outward-looking, cosmological approach. *Dark matter relationism* is epitomized in the author's use of cosmic mysteries and *relationism* to identify a DM candidate and then to confirm its validity by using this same DM candidate to provide plausible explanations for additional cosmic mysteries, including some *not known to be related to dark matter.*

In Part I of this book, a list of 14 relevant and plausible Cosmic DM Mysteries of the Universe is presented, which had been developed by the author in late 2004. These Cosmic DM Mysteries were then used to establish a list of constraints regarding the nature and characteristics of the long-sought dark matter particles. By this means, a dark matter candidate was then discovered that best conformed to the constraints established by the 14 Cosmic DM Mysteries, in conjunction with the *relationism* approach.

The author then shows in Part II of this book that this same dark matter candidate, the galaxy-orbiting relativistic proton, provides plausible explanations for the accelerating expansion of the Universe, both the "knees" and "ankle" of the cosmic ray energy distribution graph, the low star formation rates of LSB dwarf galaxies, the ignition of hydrogen fusion reactions in the first generation stars, the source of magnetic fields in spiral galaxies, how dust particles could facilitate hydrogen fusion in stars, and the four departing locations within the Local Group and Virgo Supercluster for Earthbound high-energy cosmic ray protons.

Note what was achieved. Using the relationism/cosmology approach, a dark matter *candidate* was tentatively identified. This dark matter *candidate* was then used to try to explain 14 different cosmological or astrophysical phenomena represented by the 14 Cosmic DM Mysteries initially utilized. If a dark matter *candidate* successfully provides plausible explanations for the nature and characteristics of the 14 Cosmic DM Mysteries, a significant step toward dark matter identification has been achieved.

In Part III of this book, the relativistic proton dark matter candidate is subjected to a rigorous set of 11 additional tests using 11 more Cosmic DM Mysteries. They represent 11 additional cosmological or astrophysical mysteries or unexplained phenomena reported primarily during 2005 by astronomers. Can the same dark matter *candidate* and its associated cosmology that explained the initial 14 Cosmic DM Mysteries also explain the last 11? In achieving this goal, a total of 25 Cosmic DM Mysteries may have been laid bare, thereby providing a decoding of a portion of the "DNA" of the cosmos.

Among These 25 Cosmic DM Mysteries, New And Plausible Explanations Are Provided In This Book For The Following 15 Well-Known Astrophysical Or Cosmological Phenomena That Had Not Been Adequately Explained Previously:

1. The "knees" and "ankle" of the cosmic ray proton energy distribution graph at the Earth.

2. The low star formation rates (SFRs) of low surface brightness dwarf galaxies; the higher SFRs for spiral galaxies; and the much higher SFRs for large, massive spiral galaxies for limited periods of time.

3. The ignition of hydrogen fusion reactions in the first generation stars.

4. An accelerating expansion of the Universe.

5. The source of magnetic fields in spiral galaxies.

6. How dust particles facilitate or expedite hydrogen fusion reactions in stars.

7. The empirical Schmidt law correlating star formation rate and the average molecular hydrogen density on the surfaces of isolated spiral galaxies.

8. Strongly self-interacting dark matter particles in halos of dwarf and LSB galaxies in contrast to weaker self-interacting dark matter particles in the halos of galaxy clusters, implying the existence of two types, forms, or modes of dark matter particles.

9. The extremely high star formation rates of starburst galaxies created by the merging of spiral galaxy clusters.

10. The blue and blue-white stars, as young as one million years old, in the spiral arms of mature spiral galaxies, which also contain five-billion-year-old red stars in their nuclei.

11. The process of two galaxy clusters merging that results in a source of ultra-high energy protons called ultra-high energy cosmic rays, or UHECRs.

12. The various departing locations of Earthbound cosmic ray protons with energy levels at or below 5×10^{19} eV.

13. That dark matter halos are almost spherically shaped, while their enclosed spiral galaxies are disk shaped.

14. That starburst galaxies usually exhibit new blue star formation primarily in their galaxy nuclei while spiral galaxies exhibit new blue star formation in their spiral arms.

15. The growth of the spiral galaxy disk of the Andromeda galaxy (M31) by a factor of three, by accretion.

The above phenomena were selected because they are better known by astronomers, and explanations proposed for them previously, if any, are not widely accepted. Most of these 15 phenomena are *not* normally associated with dark matter.

Why Another Dark Matter Approach Is Necessary:

Many cosmologists believe that cold dark matter seems to explain the development of the large-scale structure of the Universe better than does warm dark matter. They also believe that warm dark matter seems to explain the formation of galaxies better than does cold dark matter.

This dilemma faced by the cosmologists today probably was created by three unnecessary constraints that pioneering cosmologists apparently placed on the dark matter particles; namely, that the dark matter particles have no coulomb charge, are not influenced by magnetic fields, and cannot be transformed into normal baryonic matter.

Drexler's relativistic proton dark matter theory/cosmology avoids these three "unnecessary" constraints and provides plausible explanations for the 15 well-known, but unexplained, astrophysical or cosmological phenomena listed on the previous two pages. Drexler invites the proponents of cold and warm dark matter to offer their explanations for these 15 Cosmic DM Mysteries.

In Appendix A, the scientific community's long-held objections to any proton dark matter candidacy are presented along with the author's responses.

A 18-page glossary is provided, beginning on page 269.

Author's Simplifications:

To simplify the discussions in this book, the dark matter focus will be on the relativistic protons, even though there is one helium nucleus accompanying every 12 protons. Also, to simplify the discussions, the coulomb forces between protons are not utilized quantitatively. Further, to simplify discussions, the term galaxy *disk* is used even if the galaxy is irregular in shape.

PART I

CHAPTER 1

The Search For The Identity Of Dark Matter

Some cold dark matter (CDM) cosmologists believe that considerable progress has been made during the past 20 years.[1] They believe that the non-baryonic cold dark matter theory fits the cosmic microwave background (CMB) data, explains why the observed galaxies have specific sizes and numbers, and that CDM has been used successfully in predicting galaxy distributions in the Universe. They also believe that the CDM theory makes correct predictions about galaxy growth, mergers, and shapes of dark matter halos around galaxies and around galaxy clusters. They feel that for any new competing theory of dark matter and structure formation to be considered for adoption, it would have to compete successfully against the CDM theory. This book challenges most of these dark matter beliefs of the CDM cosmologists. CDM can be described as shown in the following section.

The Nature And Characteristics Of The Theoretical Cold Dark Matter:

- CDM particles are assumed to be uncharged and weakly interacting massive particles (WIMPs).

- CDM particles are assumed to be very slow moving.

- CDM particles are assumed to contain no protons or neutrons.

- CDM particle mass is theorized to have 35 to 10,000 times the mass of a proton at rest.

- CDM particles are assumed to follow the gravity-based laws of Newton and Kepler.

- Gravity-based CDM halo morphology would be expected to follow the shape of the enclosed galaxy.

- Bottom-up CDM theory of galaxy formation predicts that small galaxies merge to form the large ones.

- Theoretical CDM particles are assumed to be WIMPs and neutralinos (super-symmetric WIMP particles).

In order to evaluate a competing new dark matter theory in a reasonable amount of time, the author proposes that the initial research effort should lean heavily on analysis of existing astronomical data and on multi-faceted comparisons with the WIMPs and neutralinos of the CDM theory. The author believes that a combined scientific and logical approach should be employed that utilizes a group of astronomically based Cosmic DM Mysteries and the constraints they place on the nature and characteristics of dark matter to identify dark matter and to highlight the strengths and weaknesses of the competing dark matter candidates.

The Ockham's Razor Approach To Analyzing Dark Matter Candidates:

The proposed initial research direction is somewhat based upon the logical system called Ockham's razor, which was developed by the English logician, William of Ockham, in the 14th Century. Simply described, Ockham's razor logic states that a researcher should not make more assumptions than absolutely necessary in seeking explanations for observed phenomena. That is, if there are two theories proposed to explain a scientific phenomenon, the theory that requires the fewer assumptions to explain it should be favored. However, if there are two competing theories and only one astrophysical or cosmological mystery, it could be difficult to select the winning theory on the basis of that one mystery alone. In this case, additional astronomical, astrophysical, and cosmological phenomena or mysteries should be employed to facilitate the competition.

Analyzing Dark Matter Candidates By "Cosmic DM Mysteries" And "Signature Characteristics":

If one were to assemble a list of astronomically based astrophysical and cosmological phenomena that have relevance to competing dark matter theories, we might be able to subject the two dark matter theories to a pentathlon or decathlon competition in order to crown the winning theory. To utilize this approach, the author has selected a group of 14 "Cosmic DM Mysteries" along with their inherent constraints on the nature and characteristics of dark matter for a multi-faceted analysis now and for a competition with other dark matter candidates in the future. None of these 14 "Cosmic DM Mysteries" involves any aspects of the early Universe, which would of necessity involve unproven theories that are equivalent to assumptions, which must be minimized in order to be true to Ockham's razor logic.

A *"Cosmic DM Mystery"* of the Universe is an astronomically based cosmic phenomenon that is either considered a mystery or, at the least, is not well understood. Such cosmic phenomena may be facilitated, expedited, influenced by, or have a special relationship with dark matter.

Examples include dark matter halos around galaxies and galaxy clusters, the accelerating expansion of the Universe, the star rotation curves of galaxies, or the ultra-high-energy proton cosmic rays that might be influenced by a Universe filled 80% to 85% with some type of dark matter.

Each of these Cosmic DM Mysteries constrains the characteristics or nature of dark matter such as the way that Fritz Zwicky[2] did with the rotation curves of galaxies within galaxy clusters and that Vera Rubin[3,4] did with rotation curves of stars and atomic hydrogen within galaxies. That is, if a group of Cosmic DM Mysteries in conjunction with their related group of constraints is chosen to try to identify dark matter, then the "Signature Characteristics" of any considered dark matter candidate would have to be compatible with this same group of constraints.

The first goal was to select the relevant astronomically based Cosmic DM Mysteries that may be created by, influenced by, or have a special relationship with most potential dark matter candidates. Each Cosmic DM Mystery would represent a mysterious or poorly understood cosmic phenomenon in the category of an astrophysical or cosmological phenomenon.

The second goal was to utilize the Cosmic DM Mysteries and the astrophysical/cosmological constraints they impose on any dark matter candidates to screen out non-conforming candidates and to select the most logical choice for the dark matter particles.

The third goal was to provide a descriptive set of characteristics of the selected dark matter candidate in the form of a large group of Signature Characteristics. The most logical choice for the dark matter particles turned out to be *galaxy-orbiting relativistic protons that are linked to cosmic ray protons*. The author was then able to compile a list of two dozen Signature Characteristics that were plausible and compatible with galaxy-orbiting relativistic proton dark matter.

Finally, these two dozen Signature Characteristics (SigChar) of the tentatively selected dark matter candidate were utilized in trying to provide a plausible explanation for the nature of each of the Cosmic DM Mysteries and the relationships of the Cosmic DM Mysteries with the selected dark matter candidate. When the dark matter candidate provided plausible explanations for at least 10 of the 14 Cosmic DM Mysteries, it move up a notch in its level of acceptance.

This overall approach highlights the significance of the dark matter *relationism* concept in cosmology and may have yielded some possible astrophysical or cosmological discoveries.

CHAPTER 2

Additional Approaches To Dark Matter Research

Previously, dark matter research focused almost entirely on the theoretical existence of uncharged dark matter particles. The research approaches that had been used included:

1. Conducting computer simulations based upon uncharged particles.
2. Determining the ratios of dark matter to ordinary matter in galaxies and clusters.
3. Exploring the reasons for the flat rotation curves of spiral galaxies.
4. Utilizing special detectors in searching for uncharged WIMPs and neutralinos.
5. Employing gravitational lensing utilizing intervening dark matter distributions.
6. Measuring galaxy cluster motion to reveal presence of dark matter superstructure.
7. Recording and analyzing the mass density ripples in the early Universe via the cosmic microwave background (CMB) through the Cosmic Background Explorer (COBE) and the Wilkinson Microwave Anisotropy Probe (WMAP).

Since dark matter is responsible for about 80% to 85% of the mass of the Universe, it is likely to influence or have relationships with a number of Cosmic DM Mysteries that have not been adequately utilized or explored in prior research. The development of additional research approaches to uncover the nature of dark matter was guided by the following three related hypotheses:

1. The Universe comprises Cosmic DM Mysteries formed into a system arrangement in which some of the Cosmic DM Mysteries may be mutually related.

2. Some of the principal mysteries of astrophysical cosmology may be linked, making it possible to find a single unique astrophysical theory to solve several of the cosmological mysteries.

3. Since traditional cosmology and astrophysics have failed in identifying dark matter, the source of high- and ultra-high energy cosmic rays, the accelerating expansion of the cosmos, and how galaxies are formed, perhaps an expanded approach to dark matter research should be tried. That is, in researching the nature of dark matter, an analysis of its influence on and its relationships with relevant Cosmic DM Mysteries of the Universe might provide important clues as to some of the characteristics of dark matter. This is the analytical approach that the author has designated *relationism*.

CHAPTER 3

Dark Matter Research Guided By The Three Related Hypotheses

The dark matter research approaches utilized by the author were guided by the three related hypotheses discussed in the previous chapter. The approaches included the selection of the Cosmic DM Mysteries of the Universe that possibly could be strongly influenced by or have a relationship with the enormous mass of dark matter particles in the Universe.

The potentially relevant Cosmic DM Mysteries that were hypothesized to be influenced by the multitudinous dark matter particles included the accelerating expansion of the Universe, new star ignition, the spherical dark matter halos around spiral galaxies, the extragalactic magnetic fields, the flat rotation curves of spiral galaxies, the linearly rising rotation curves of LSB dwarf galaxies, and the ratio of dark matter to ordinary matter in the Universe.

The next step after selecting a group of relevant Cosmic DM Mysteries was to try to identify the type of dark matter particles that could create, cause, form, lead to, influence, or have a relationship with many or most of the selected Cosmic DM Mysteries. An evolving series of rhetorical questions representing Cosmic DM Mysteries was considered along the way, leading to the following list of 14 such questions.

What Type Of Dark Matter (DM) Particles Could:

1. Form spherical dark matter halos around spiral galaxies and dark matter halos around galaxy clusters?

2. Cause the accelerating expansion of the Universe and possibly store dark energy?

3. Be transformed into low-velocity hydrogen, protons or proton cosmic rays?

4. Create the magnetic fields within and around spiral galaxies?

5. Be concentrated in the long, large filaments of dark matter (announced by NASA/Harvard on September 8 and 9, 2004; see Chapter 33), which form galaxy clusters where two dark matter filaments intersect?

6. Create large, mature spiral galaxies less than 2.5 billion years after the Big Bang?

7. Create spherical dark matter halos having predictable outer and "hollow" core diameters?

8. Provide angular momentum to spiral galaxies and dark matter halos?

9. Create galaxies without a central dark matter density cusp?

10. Create a starless galaxy or a low surface brightness (LSB) dwarf galaxy with low star formation rates?

11. Lead to linearly rising rotation curves for LSB dwarf galaxies and to flat rotation curves for spiral galaxies?

12. Form about 80% to 85% of the mass of the Universe, the remainder being hydrogen, helium, etc.?

13. Ignite hydrogen fusion reactions of second-generation stars utilizing hydrogen molecules, helium, and dust and ignite fusion reactions of the first-generation stars with only hydrogen and helium atoms?

14. Create the first "knee" at 3×10^{15} eV, the second "knee" between 10^{17} eV and 10^{18} eV, and the ankle at 3×10^{18} eV of the cosmic ray energy distribution near the Earth?

It appears that there is at least one dark matter candidate that could have an influence on or relationship with at least 10 of the above-indicated Cosmic DM Mysteries -- namely, *galaxy-orbiting relativistic protons*. They are constrained by the galactic and extragalactic magnetic fields into Larmor Radius spiral orbits (see Chapter 6) within galaxies and within dark matter halos around galaxies and around galaxy clusters. They also could form long, large filaments of dark matter. When they lose kinetic energy, they become cosmic ray protons bombarding star systems. The next paragraphs are about the nature and characteristics of the galaxy-orbiting relativistic proton dark matter.

Some Laws And Principles Of Physics And Astrophysics Relevant To Relativistic Protons:

- Astrophysical dynamo effect -- Relativistic protons in Larmor orbits create galactic magnetic fields through the astrophysical dynamo effect. That is, the proton paths determine the magnetic fields and the magnetic fields determine the proton paths, after an emergent evolution period involving millions to billions of years.

- Baryonic limit to dark matter -- There are no constraints regarding dark matter that preclude the possibility that the number of relativistic protons in dark matter could total as much as 15% of the total number of baryons in the Universe. (The total mass of the relativistic protons still could be a factor of six times greater than the non-relativistic baryons in the Universe.)

- Law of Conservation of Linear Momentum -- The total linear momentum of a group of linearly moving objects must remain constant. Therefore, if the combined mass of a moving galaxy cluster and its dark matter halo declines, its linear velocity should increase.

- Larmor Radius equation (see Chapter 6) -- This equation determines the radii of curvature of the paths of relativistic protons moving through orthogonal magnetic fields. The Larmor Radius of such paths increases directly with proton kinetic energy and inversely with the orthogonal magnetic field strength.

- Relativistic proton synchrotron radiation -- This radiation of photons increases directly with the orthogonal magnetic field strength and inversely with proton kinetic energy. (Proton synchrotron radiation is lower than electron synchrotron radiation by a factor of 11 trillion.)

- Einstein's Special Theory of Relativity -- This explains that a proton traveling near the speed of light could have a relativistic mass that is many orders of magnitude greater than the mass of a proton at rest. Therefore, if the average relativistic mass of the dark matter protons is about 50 times the rest mass of a proton and the number of relativistic dark matter protons in the Universe is, say, about 12% of the total number of baryons in the Universe, then the total relativistic dark matter mass would be about six times the mass of the ordinary non-relativistic baryonic matter in the Universe. This matches scientists' estimates for the ratio of dark matter mass to total ordinary matter mass in the Universe.

In addition to the above, *astrophysical emergence* or *emergent evolution* could cause relativistic protons to evolve into complex organizational groupings, leading to the formation of dark matter halos around galaxies and around galaxy clusters. Emergent evolution (also known as emergence, astrophysical emergence, or collective self-organization) is a theory that new characteristics and qualities appear in the evolutionary process at larger, more complex organizational levels that cannot be predicted by studying less complex levels of organization. The large, complex organizational levels are determined by a rearrangement of the pre-existing entities.

Also, for emergent evolution principles to succeed, the pre-existing entities must be strongly self interacting. That is, in order for emergent evolution behavior to manifest itself, the pre-existing entities must interact strongly with each other according to the laws of physics. Relativistic protons do interact strongly with each other, but the *weakly interacting* massive particles of the cold dark matter theory do not.

Galaxy-orbiting relativistic protons, posited by the author to be the long-sought dark matter particles, are expected to have the following Signature Characteristics (SigChar), denoted as SigChar A through X, as described in the chapters that follow.

CHAPTER 4

SigChar A

Dark Matter Proton Energies

Relativistic protons in dark matter halos would have energies ranging from about 10^{16} eV to 5×10^{19} eV.[5] (See the cosmic ray energy distribution graph in Appendix B, Slide #17.) Their energies would decline continuously through synchrotron radiation losses, leading to a proton flow radially inward into the higher magnetic field of the enclosed galaxy.

In the case of the Milky Way, the galactic magnetic field is about 2,000 times greater than the extragalactic magnetic field. Therefore, protons entering the Milky Way galaxy would experience rising synchrotron radiation energy losses and declining kinetic energies, with the result that the kinetic energies of most of the protons entering a star system or the solar system as cosmic ray protons could be reduced significantly below their galaxy-orbiting dark matter halo levels.

Note that the galaxy-orbiting relativistic protons in a galaxy halo are the dark matter protons, and their flux would be many orders of magnitude greater than the flux of the cosmic ray protons entering all the star systems of the same galaxy within one year.

Also note that, although there is a strong limit from primordial nucleosynthesis on the maximum baryonic *particle density* in the Universe, the high relativistic mass of the dark matter relativistic protons provides the necessary dark matter *mass density* while staying within the low baryonic *particle density* limits.

CHAPTER 5

SigChar B

The Milky Way's Magnetic Fields

The dark matter protons move through the extragalactic (intergalactic) magnetic field of about $10^{(-9)}$ gauss in the dark matter halo of the Milky Way and through the 2,000 times higher magnetic field of the Milky Way of about $2 \times 10^{(-6)}$ gauss.[6]

The creation of magnetic fields surrounding spiral galaxies requires the flow of coulomb charges through space. Relativistic dark matter protons orbiting galaxies will create such magnetic fields through the astrophysical dynamo effect.

In the astrophysical dynamo effect, relativistic protons in Larmor orbits create galactic and extragalactic magnetic fields. That is, the proton paths determine the long-term magnetic fields and the magnetic fields determine the long-term proton paths, after an *emergent evolution* period involving millions to billions of years. Electrons do not seem to play an important role in creating astrophysical magnetic fields since they lose their energy too rapidly through synchrotron radiation losses and, therefore, do not create large-magnitude steady-state coulomb charge flows.

CHAPTER 6

SigChar C

Larmor Radius Equation

The dark matter protons' spiral orbiting paths in dark matter halos are determined by the Larmor Radius equation,[6] as follows:

$$r = 110 \text{ Kpc} \times \frac{10^{-8} \text{ gauss}}{B} \times \frac{E}{10^{18} \text{ eV}}$$

where **Kpc** means kilo parsec and one parsec equals 3.26 light-years and where **B** is the orthogonal magnetic field.

The Larmor Radius for a 10^{16} eV proton in the Milky Way halo's extragalactic magnetic field of 10^{-9} gauss is 11 Kpc; for a 10^{17} eV proton, it is 110 Kpc; and for a 10^{18} eV proton, it is 1,100 Kpc.

The diameter of the Milky Way galaxy is about 100,000 light-years, or 30.7 Kpc, and its radius is about 15 Kpc. Astronomical observations have found that the dark matter halo around a spiral galaxy extends about 10 to 20 times the size of the galaxy's visible regions. Using a factor of 15, the radius of the dark matter halo around the Milky Way would extend to perhaps 225 Kpc.

Thus, some 10^{16} eV dark matter protons with a Larmor Radius of 11 Kpc would penetrate the Milky Way's surface that is located at a radius of 15 Kpc. A 10^{17} eV proton with a Larmor Radius of 110 Kpc would remain within the DM halo's 225 Kpc radius, but a 10^{18} eV proton probably would find itself eventually within the DM halo surrounding the Local Group cluster of galaxies, whose proton energies probably range between $3x10^{17}$ eV and $6x10^{18}$ eV.

Gravitational accelerations from the enclosed galaxy mass onto the dark matter halo should be relatively negligible since gravitational accelerations are many orders of magnitude smaller than electromagnetic accelerations that establish the protons' Larmor radii orbital paths.

This Larmor-Radius-based analysis seems to provide additional evidence that relativistic protons are a credible dark matter candidate. No other currently proposed dark matter candidate can be used to estimate the size of the Milky Way's DM halo or the size of the Milky Way itself.

CHAPTER 7

SigChar D

Milky Way's Dark Matter Halos And Proton Energies

SigChars A through C plus the diameters of the Milky Way, its dark matter halo, and its Local Group's halo could lead to a rough estimate that relativistic protons in the Milky Way's dark matter halo probably would have energies of about 1×10^{16} eV at its inner core diameter and, from the Larmor Radius equation, about 2×10^{17} eV at its outer diameter.

The dark matter halo of its Local Group galaxy cluster probably would have protons with energies of about 3×10^{17} eV at its inner core diameter and probably near 6×10^{18} eV at its outer diameter. These estimated proton energies are about 30 times greater than those of the Milky Way's DM halo. The dark matter halos in and around the Virgo Supercluster probably would contain UHE protons with energies above 6×10^{18} eV.

In the well-known "leaky box" model for cosmic rays, higher energy protons escape from a galaxy into its halo. In contrast, in this model of a hierarchy of dark matter halos enclosing smaller dark matter halos, a model is employed wherein relativistic protons that have lost kinetic energy through synchrotron radiation losses or collisions with particles or photons, leak into the next-smaller, lower energy halos and finally into the star systems of a galaxy as *cosmic ray protons*.

Therefore, most of the cosmic ray protons arriving at the solar system would tend to be the lowest energy relativistic protons previously residing in the Milky Way, in its dark matter halo, and in the dark matter halos of the Local Group and the Virgo Supercluster. Also, under this theory, the higher the energy of the arriving cosmic ray protons, the greater the distance they probably have traveled to reach the solar system. Thus, the fall-off of proton flux with proton energy of the Earth-arriving cosmic ray protons could be more rapid than that for the dark matter protons currently remaining in the three halos and in the galaxy. See Chapter 42 (Cosmic DM Mystery #14) for the estimated diameters of the Milky Way, its halo, etc.[5,7,8]

CHAPTER 8

SigChar E

Paths Of Protons

The orbits and paths of relativistic protons would be determined by their kinetic energies, the local orthogonal magnetic field strengths, and the Larmor Radius equation. Their gravitational accelerations from nearby mass objects should be relatively negligible because gravitational accelerations are many orders of magnitude smaller than electromagnetic accelerations.[6] Synchrotron radiation losses would continuously reduce the radii of the proton paths, causing dark matter protons to move from a large diameter dark matter halo into its enclosed galaxy and eventually into its star systems as cosmic ray protons (or as hydrogen, as will be explained later) over millions to billions of years.

If the relativistic protons lose energy in collisions with dust, molecules, or photons, the radii of their paths will also be reduced.

CHAPTER 9

SigChar F

Proton Streams Creating Magnetic Fields

The dark matter halos surrounding spiral galaxies, even very young galaxies, exhibit magnetic fields. The relativistic protons of the author's dark matter theory follow spiral orbiting paths that are determined by a halo's orthogonal magnetic field strength. These orbiting protons also participate in creating and maintaining the halo's magnetic field strength via the astrophysical dynamo effect, as described in *"The Magnetic Universe -- Geophysical and Astrophysical Dynamo Theory."* [9] Also see the A&A paper entitled, "Strong magnetic fields and cosmic rays in very young galaxies"[10] and a related paper, astro-ph/0402662.[11]

Other researchers reported on March 9, 2005, "There are several viable alternatives to explain the coherent magnetic fields that we observe [in the Large Magellanic Cloud (LMC). Potentially most pertinent for the LMC is the cosmic-ray driven [astrophysical] dynamo ...."[12,13]

CHAPTER 10

SigChar G

Proton Flux And Kinetic Energy In Halos

An analysis of SigChars A through F (particularly SigChar D) and the knowledge that cosmic ray proton flux density in the Universe declines rapidly with proton kinetic energy, lead to the conclusion that a galaxy's dark matter halo would be expected to have the highest proton flux density and lowest kinetic energy protons near its inner core (near the galactic magnetic field) and the lowest proton flux density and highest energy protons near its outer diameter (in the extragalactic/intergalactic magnetic field).[5,6]

Therefore, the protons leaving the dark matter halo to enter the enclosed galaxy are the lowest energy relativistic protons from the halo. They will eventually play an important role in star formation. When two galaxy clusters are merging, their magnetic fields combine and become distorted, and much higher energy DM halo protons could plunge into the enclosed galaxies, thereby dramatically increasing the star formation rate by more than an order of magnitude, as in the case of the well-known starburst galaxies.

CHAPTER 11

SigChar H

Proton Relativistic Mass Losses From Synchrotron Radiation

Whenever relativistic protons move across an orthogonal magnetic field, they are deflected (accelerated) and radiate photons as synchrotron radiation. The smaller the radius of curvature of the spiral path of a proton, the more photon energy it emits as synchrotron radiation, which causes the proton's kinetic energy, velocity, and relativistic mass to decline. Thus, a galaxy-orbiting stream of dark matter halo protons with declining relativistic mass would cause the gravitational acceleration on nearby galaxies or gas clouds to decline.[14] This same decreasing relativistic mass effect also would be experienced by galaxy clusters and their surrounding dark matter halos.

(Note that a proton's kinetic energy losses from synchrotron radiation are lower by a factor of about 11 trillion compared to that of an electron. Therefore, dark matter relativistic protons could remain in dark matter halos for billions of years, provided their collisions with photons, dust, hydrogen, or helium are not substantial.)

CHAPTER 12

SigChar I

Magnetic Bulges Leading To Increased Synchrotron Radiation From Protons

Whenever a magnetically constrained relativistic proton encounters a rise in the orthogonal magnetic field, its acceleration perpendicular to the proton path direction increases. This causes its synchrotron radiation energy losses to increase; and as a result, its velocity, relativistic mass, and radius of curvature of its spiral path decline.[14,15]

More specifically, when an ultra-high energy (UHE) proton moving along a certain path crosses orthogonal magnetic field lines, it is deflected depending upon the magnetic field direction and field strength. Any magnetic deflection of a UHE proton reduces its velocity in the direction of its original path for two reasons: The deflection itself will cause the proton's direction to change, and the increased synchrotron radiation energy losses will reduce its kinetic energy and its velocity.

Thus, if UHE protons that are moving through space at a certain velocity encounter a magnetic bulge (increase in the magnetic field strength), they will not pass through the magnetic bulge region as quickly as through a no-magnetic-bulge region.

CHAPTER 13

SigChar J

Why Dark Matter Halo Protons Enter Their Enclosed Galaxy And Lose Relativistic Mass

It follows from the previous chapter that a stream of magnetically constrained relativistic protons in a dark matter halo skimming the surface of its enclosed galaxy would experience the same magnetic-bulge effect. This would be due to the higher orthogonal magnetic field of the enclosed galaxy, thereby causing the orbiting relativistic proton stream to move deeper into the enclosed galaxy.

This, in turn, leads to increased synchrotron radiation and a loss of proton kinetic energy and relativistic mass. By this means, the combined mass of the galaxy and its dark matter halo would decline. (A similar process can occur if the halo protons collide with dust, photons, or hydrogen/helium clouds causing the protons to slow down, thus decreasing their path radius and increasing their synchrotron radiation energy losses.)[14,15]

CHAPTER 14

SigChar K

Protons/Helium Nuclei Collisions With Hydrogen Clouds

High-velocity collisions by the relativistic protons (and by the accompanying relativistic helium nuclei) with atomic and molecular hydrogen in compressed interstellar hydrogen clouds in a galaxy could lead to a hydrogen fusion reaction and the ignition of new stars. Note that colliding protons with energies of 10^{15} eV would be 1,000 times more powerful than those produced by today's man-made accelerators; and, therefore, 10^{15} eV protons could play a significant role in initiating and maintaining hydrogen fusion nuclear reactions.[16]

Such collisions will also cause the creation of muons that can catalyze hydrogen fusion reactions in stars, as described in Chapters 26 and 27.

CHAPTER 15

SigChar L

Linearly Rising Rotation Curves Indicating LSB Dwarf Galaxy DM Halos Are "Weakly Centrally Concentrated" (i.e., "Hollow")

A paper on LSB dwarf galaxy dark matter halos, authored by J. Bailin et al and published in MNRAS 14 February 2005,[17] contains some pertinent statements:

> These studies focused on the shape of the LSB galaxy rotation curves, which rise approximately linearly with radius in a manner one expects if the mass density profile contained only a shallow central cusp or a constant density core. ... Thus, a slowly rising rotation curve could be interpreted as indicative of a halo that is weakly centrally concentrated.

An excerpt from the abstract reads:

> We find that our LSB galaxy analogues occupy haloes that have lower [mass] concentrations than might be expected ....

SigChar T in Chapter 23 provides a theory for the LSB dwarf galaxies' low star production rates and, in the process, translates the expression "weakly centrally concentrated"

LSB dark matter *cored* halo into the simpler term "hollow" LSB dark matter halo. (Note: A galaxy rotation curve plots the orbital velocities of its stars and hydrogen gas as a function of their radial distances from the nucleus of the galaxy and into the surrounding dark matter halo.)

There are well over 100 astro-ph papers on the subject of *cored* dark matter halos/haloes. For example, see references *17* and *55*.

Flat Rotation Curves Of Spiral Galaxies:

The well-known flat galaxy rotation curves discovered by Vera Rubin, involving the stars and hydrogen in spiral galaxies, indicate that the spherical mass of dark matter contained within a sphere of radius r from the nucleus of the spiral galaxy, increases linearly with radius r through the galaxy and into the halo.[3,4] This means that the dark matter mass density at radius r must be declining approximately as the square of the radius r. It is important to note that this astronomically determined feature of dark matter halo mass is inherent and distinctive in Drexler's relativistic proton dark matter halo, as is explained in the following paragraphs.

Relativistic proton streams in dark matter halos, as pointed out in SigChar G in Chapter 10, are expected to have a rapidly declining proton flux density as a function of proton path radius. That is, the highest proton flux density would be near the "hollow" inner core of the dark matter halo, and the lowest proton flux density would be near the dark matter halo's outer diameter.

More specifically, while the *particle* flux density of the cosmic ray relativistic protons falls as a power law of their energies, with the exponential decline ranging between 2.7 and 3, as determined from the cosmic-ray energy distribution graph[5,18] (see Appendix B, Slide #17), the relativistic mass of each proton rises linearly with kinetic energy and is highest at the dark matter halo's outer diameter.[5]

This leads to the approximation, according the relativistic proton dark matter theory/cosmology, that the *mass* flux density of the orbiting protons within the galaxy and within its halo, falls nearly inversely as the square of the radius r from the nucleus of a spiral galaxy outward, thereby approximately satisfying Vera Rubin's flat-rotation-curves requirement.

Note that Vera Rubin's calculated dark matter mass distribution in a spiral galaxy and its halo, from her astronomically derived flat galaxy rotation curves, roughly matches the mass distribution within a spiral galaxy and its halo according to the relativistic proton dark matter theory/cosmology.

This compatibility between the flux/energy spatial distribution of the cosmic ray protons in the Milky Way and Vera Rubin's rotation curve requirements appears to provide additional evidence supporting the relativistic proton dark matter theory/cosmology.

CHAPTER 16

SigChar M

An Explanation For The Two "Knees" And "Ankle" Of The Cosmic Ray Energy Distribution

The relativistic proton dark matter theory/cosmology may be able to explain the observed energies of the two "knees" of the cosmic-ray energy distribution (near Earth) graph at about 3×10^{15} eV and at about 3×10^{17} eV and at the "ankle" of the graph at 3×10^{18} eV in terms of four different departing locations for Earthbound cosmic ray protons. See Chapter 7 (SigChar D) and related references *5, 7, 8, 18*, and also Appendix B, Slide #17.

1. The departing location of most of the cosmic ray protons approaching the Earth with energies between 10^9 eV and under 1×10^{16} eV is probably the Milky Way, since the relatively strong galactic magnetic field confines these lower energy cosmic ray protons to the galaxy. (The "hollow" core of the dark matter halo of the Milky Way was created at an earlier time, when the Milky Way was much smaller. At that time, orbiting DM protons with energies much less than 3×10^{15} eV would lose a significant portion of their kinetic energy through

synchrotron radiation losses and would leave the core region of the DM halo, thereby enlarging the "hollow" core.)

2. Most of the Earth-arriving cosmic ray protons with departing kinetic energies between 1×10^{16} eV and about 2×10^{17} eV to 3×10^{17} eV probably come via the Milky Way's dark matter halo through the Milky Way to the solar system, for reasons explained in Chapter 13 (SigChar J). Apparently, the galactic magnetic field is too weak to prevent these high energy cosmic ray protons from reaching the solar system.

3. Most of the cosmic ray protons with departing energies between 3×10^{17} eV and 6×10^{18} eV probably arrive at the Earth via the dark matter halo surrounding the Local Group galaxy cluster for reasons similar to those in Chapter 13 (SigChar J) and apparently because the intervening magnetic fields are too weak to prevent these very high energy protons from passing through the galaxy cluster, the Milky Way's dark matter halo, and the Milky Way before finally reaching the solar system.

4. The dark matter halos in and around the Virgo Supercluster probably would contain UHE protons with departing energies above 6×10^{18} eV and probably would be the departing location of the highest energy cosmic ray protons arriving at the Earth. Note that at 10^{19} eV, only 3 to 4 protons per square kilometer arrive at the Earth each century.

CHAPTER 17

SigChar N

Proton Synchrotron Radiation Losses And Proton Collision Losses Possibly Could Lead To An Accelerating Expansion Of The Universe

The author's relativistic proton dark matter theory leads to a possible explanation of an accelerating expansion of the Universe without any additional assumptions.[14,19] This theory posits that the observed dark matter halos around spiral galaxies and their clusters may be comprised of relativistic-velocity/relativistic-mass protons following halo-size spiral/orbital paths through the dark matter halos as they continually lose relativistic mass due to kinetic energy losses from synchrotron radiation and from collisions with photons, dust, and atoms of hydrogen and helium.

The well-known galactic and extragalactic magnetic fields would establish both the halo-size Larmor Radius spiral/orbital paths of the high-relativistic-mass protons and their relativistic mass losses due to the synchrotron radiation losses. With galaxy *clusters* already experiencing separation velocities according to the Hubble Law, the decreasing relativistic masses of the dark matter halos around galaxy clusters and around the clusters' galaxies should cause the galaxy cluster separation velocities to increase under the Law of Conservation of Linear Momentum, thereby leading to an accelerating expansion of the Universe.

The energy to accomplish this separation acceleration perhaps could be derived directly from the conversion to kinetic energy of some of the relativistic mass of the dark matter halos of the galaxy clusters and the dark matter halos of the galaxies in the clusters. In this case, the long-sought mysterious *dark energy* might be related to the total kinetic energy and/or the energy equivalent of the relativistic mass of all the dark matter protons in the various galaxies, galaxy clusters, and dark matter halos.

However, for the earlier, smaller, and denser Universe of more than six billion years ago, the smaller distances between galaxy clusters may have caused the gravitational attraction between them to be high because of the inverse square of those then-smaller distances between galaxy clusters. This inverse-square law relationship may be a principal reason that the accelerated expansion did not begin until about six billion years ago when the Conservation of Linear Momentum effect and the declining galaxy cluster mass finally overcame the gravitational attraction between galaxy clusters, which were closer together in the earlier epoch.

Note that in this theory of an accelerating expansion of the Universe, the concept of "dark energy," as such, does not play a principal role. Also, the accelerating expansion theory relies on the validity of the relativistic proton dark matter theory/cosmology, but this relativistic dark matter theory does not rely upon the validity of the theory of an accelerating expansion of the Universe.

CHAPTER 18

SigChar O

Radiating DM Halo Protons Become Cosmic Ray Protons

When relativistic protons in a dark matter halo give up some kinetic energy through synchrotron radiation or through collisions with dust, photons, molecules, or atoms of hydrogen or helium, they decelerate into slower moving relativistic protons. For example, when the kinetic energy of some of the relativistic protons in the Milky Way's dark matter halo, extending into the galaxy, is reduced below 3×10^{15} eV, the relativistic protons' synchrotron radiation losses increase and their kinetic energy, relativistic mass, and radius of curvature of their spiral orbits decline. The relativistic dark matter protons then move deeper into the galaxy and eventually plunge into star systems as cosmic ray protons.[14] Under the dark matter theory/cosmology, most of the higher energy Earthbound cosmic ray protons would depart from the Local Group galaxy cluster.

CHAPTER 19

SigChar P

Long, Large DM Filaments Creating Galaxy Clusters

The September 8 and 9, 2004 news releases from Harvard and NASA entitled, "Motions in nearby galaxy cluster reveal presence of hidden superstructure" (see Chapter 33) regarding Chandra x-ray images of the Fornax cluster, states:

> Astronomers think that most of the matter in the universe is concentrated in long, large filaments of dark matter and that galaxy clusters are formed where these filaments intersect.

This relatively new top-down theory of galaxy cluster formation is compatible with the relativistic proton dark matter theory as described in the author's book published in December 2003.⁵

Prior to the September 2004 news releases, the theory of cold dark matter galaxy formation was based upon the bottom-up hierarchical model wherein small galaxies form first and then gravitationally move together over time to form larger galaxies and galaxy clusters.

CHAPTER 20

SigChar Q

Mature Galaxies In A Young Universe

The recent discovery of the existence of mature red galaxies only about 2.5 billion years after the Big Bang[20,21,22] (and confirmed by the Carnegie Observatories on March 10, 2005) can be explained using the relativistic proton dark matter theory. Relativistic protons from the cores of the DM halos form proto-galaxies. They also collide with photons and hydrogen atoms, thereby creating muons and electrons with which the protons combine to form atomic hydrogen. This formation of atomic hydrogen and muons within the proto-galaxies, followed by bombardment by relativistic protons from the DM halo, leads to the formation of both molecular hydrogen and stars. (See Chapter 26.)

On the other hand, this recent discovery of the early mature red galaxies raises questions about the cold dark matter bottom-up theory of galaxy formation, which involves only slow-moving, *non-baryonic* WIMP particles that would not be expected to form the early, mature, *baryonic* galaxies.

The above references *20* and *21* are articles in the July 2004 issue of Nature, entitled, "A high abundance of massive galaxies 3 - 6 billion years after the Big Bang" and "Old galaxies in the young Universe." The Carnegie Observatories had announced in a news release on March 10, 2005 that:

> Astronomers have found distant red galaxies -- very massive and old -- in the universe when it was only 2.5 billion years post Big Bang. Bang.

This quoted sentence would imply that the red galaxies were probably born as blue galaxies perhaps about 1.5 billion years post Big Bang. It is unlikely that CDM WIMPs could travel fast enough to accomplish this in that early epoch.

Also supporting the relativistic proton dark matter theory over the cold dark matter theory is the discovery announced on December 1, 2005 (see Chapter 51) of 800 young galaxies located 12.5 billion light years away that were born less than 1.2 billion years after the Big Bang.

CHAPTER 21

SigChar R

Conservation Of Angular Momentum

All spiral galaxies and their dark matter halos exhibit an angular momentum. This characteristic is consistent with the relativistic proton dark matter model in which protons move in halo-size spiral orbits around galaxies and through the galactic and extragalactic magnetic fields according to the Larmor Radius equation. This orbiting stream of relativistic protons possesses a considerable amount of angular momentum to transfer to spiral galaxies and their dark matter halos under the Law of Conservation of Angular Momentum.

The cold dark matter proponents among the world's astrophysicists and cosmologists have not explained the roles of WIMPs in providing angular momentum to spiral galaxies and their dark matter halos.

CHAPTER 22

SigChar S

No DM Cusps In The Nuclei Of Spiral Galaxies

Although the cold dark matter theory predicts dark matter cusps in the nuclei of spiral galaxies, astronomers have not detected them. The relativistic proton dark matter theory/cosmology posits that there should be no dark matter cusps at the nuclei of galaxies.

The galactic magnetic field of a spiral galaxy would tend to constrain relativistic dark matter protons from moving toward the nucleus of a spiral galaxy. When these dark matter protons lose kinetic energy and move into the galaxy as cosmic ray protons, they will encounter star systems magnetically and probably will collide with one of them before reaching the galaxy nucleus.

The Sun, a typical star, has a magnetic field strength roughly on the order of one gauss, and that of the Earth is about 0.5

gauss. Thus, the interplanetary magnetic field strength would be roughly of the order of 0.5 gauss. The synchrotron radiation losses for a relativistic cosmic ray proton passing near a star could be of the order of 200,000 times greater than that for interstellar space. Also, a star's magnetic field would tend to trap a proton passing nearby.

Therefore, relativistic cosmic ray protons should not tend to form a dark matter density cusp at the nucleus of a galaxy.

In summary, no dark matter mass density cusps are found in the nuclei of spiral galaxies. However, the cold dark matter theory predicts the formation of DM mass density cusps at the centers of spiral galaxies. The lack of DM mass density cusps has raised questions about the validity of the CDM theory.

CHAPTER 23

SigChar T

Explanations For LSB Dwarf Galaxies' Low Star Formation Rates (SFRs) And For Massive Galaxies' Very High SFRs

If the diameter of an LSB dwarf galaxy disk is smaller than the "hollow" core of its dark matter halo, the number of dark matter halo protons entering the galaxy to ignite or feed stars would be low or very low. This would occur because the dark matter protons near the DM halo's core, being farther away from the galaxy disk, would not be subjected to the full magnitude of the galaxy's much higher magnetic field strength. Therefore, the dark matter halo protons would not experience large synchrotron radiation losses and their flow into the enclosed LSB dwarf galaxy disk would be minimal. See Chapter 15 (SigChar L) and references *4* and *17*.

The author believes that LSB dwarf galaxies and starless dark galaxies probably have this smaller-galaxy-disk/larger-halo-core relationship.

The author also believes that an LSB dwarf galaxy or starless galaxy could evolve into a star-creating galaxy as the galaxy disk grows in size, over time, through hydrogen accretion until the disk becomes equal to or larger than the "hollow" core of its DM halo. This hydrogen would have been formed by slowed protons linking to electrons created by decaying muons, formed microseconds earlier by proton collisions with dust, photons, and/or hydrogen molecules.

The author further believes that the Milky Way is of this overlapping disk-halo type because it is generally believed that the dark matter mass penetrating into the Milky Way is of the same order of magnitude as the ordinary matter mass of the Milky Way, and that our galaxy is a confirmed star-forming galaxy with a medium level SFR. See Chapter 7, SigChar D.

An unusually large, massive galaxy could significantly overlap the dark matter halo's "hollow" core, leading to a very high SFR. Such a massive galaxy's disk would be reaching deep into the dark matter halo, thereby interacting with higher energy relativistic DM protons and thus creating a very active high star formation region. However, if this

SFR is too high, the replenishment hydrogen being supplied by the dark matter halo and the galaxy could be insufficient to maintain the SFR, which could fall rapidly to lower levels after a period of time. For the Milky Way and LSB dwarf galaxies, the replenishment hydrogen could be adequate to maintain their SFRs for much longer periods of time than for large, massive galaxies.

One well-known LSB dwarf galaxy is DDO154, also known as NGC 4789A and UGC8024, which contains a very large amount of atomic hydrogen gas and has a very large ratio of dark matter to ordinary matter but, for some reasons, has a very low star formation rate. Another LSB dwarf galaxy of this type with a huge disk of rotating atomic hydrogen gas, UGC 5288, was studied with a radio telescope by Liese van Zee of Indiana University using the NSF VLA telescope and reported on in a press release dated January 10, 2005, entitled, "Dwarf Galaxy Gives Giant Surprise."

Another possible reason these two dwarf galaxies have low SFRs is that they may have a shortage of molecular hydrogen, even though they have adequate amounts of atomic hydrogen. The empirical Schmidt law illustrates the importance of molecular hydrogen over atomic hydrogen in SFRs, as described in Chapter 53. In LSB dwarf galaxies with small disks, fewer and lower-energy relativistic DM protons would be bombarding the hydrogen atoms, thereby lowering the hydrogen ionization rate and the hydrogen molecule forming rate in these galaxies. On the other hand, in spiral galaxies the atomic hydrogen may be converted much faster into molecular hydrogen through the ionization of as much as 50% of the atomic hydrogen by the bombardment with high energy relativistic dark matter protons.

As will be explained in subsequent chapters, collisions of the relativistic protons with large collision cross section molecules will generate large numbers of muons, which can form muonic ions with molecules of hydrogen and helium that can catalyze hydrogen fusion reactions and thus create stars.

On February 18, 2005, Astronomy magazine published an article by Ken Croswell entitled, "The first dark galaxy?" in which astronomers from Britain (Cardiff University in Wales), Australia, France, and Italy stated:

A [rotating] cloud of gas in the Virgo cluster may be the first dark galaxy ever found. ... The mysterious object has one-tenth the Milky Way's mass but consists of hydrogen gas and dark matter -- with no detectable stars.

Yet, its mass-to-blue-light ratio is at least 10 times that of the Milky Way. The researchers' astro-ph/0502312 paper is entitled, "A Dark Hydrogen Cloud in the Virgo Cluster."[23]

On December 1, 2004, the University of Virginia announced the discovery of a galaxy, named I Zwicky 18, that existed as a galaxy in an embryonic state as a cold gas cloud of hydrogen for billions of years and "went through a sudden first starburst only about 500 million years ago." The news release is entitled, "Hubble Uncovers a Baby Galaxy in a Grown-Up Universe." The researchers' astro-ph/0408391 paper is entitled, "Deep Hubble Space Telescope/ACS Observations of I Zw 18: a Young Galaxy in Formation."[24]

It is possible that the disk of this galaxy was smaller in diameter than the "hollow" core of its DM halo for billions of years until they finally overlapped after the galaxy disk grew in size through accretion of hydrogen and helium derived from its DM halo during those billions of years.

CHAPTER 24

SigChar U

The Relativistic Energy Of All The Protons In The Universe May Provide The Energy For An Accelerating Expansion Of The Universe

Professor Pierre Sokolsky, in his 2004 book entitled, "Introduction to Ultrahigh Energy Cosmic Ray Physics"[25] wrote the following about cosmic rays in the Universe:

> If the energy density that we observe on Earth is similar to what exists in extragalactic space, a significant component of the total energy of the Universe is in cosmic rays. The cosmic ray energy density integrated over all energies turns out to be approximately 1eV/cm3. For comparison, starlight has an energy density of 0.6 eV/cm3 and the energy density of the galactic magnetic field is 0.2eV/cm3. It is clear that cosmic rays form a major constituent of the interstellar medium.

Note that the cosmic ray energy density of 1ev/cm3 only represents the cosmic ray protons and helium nuclei raining on the star systems of galaxies. The relativistic dark matter protons and helium nuclei orbiting these same galaxies should represent a very much greater amount of energy in the form of relativistic mass/energy. This mass/energy may be the energy that drives the accelerating expansion of the galaxy clusters, as described in Chapter 17 (SigChar N).

These proton Larmor Radius orbits also create magnetic fields under the astrophysical dynamo effect that store a significant amount of magnetic energy. See references *9, 10,* and *11.*

CHAPTER 25

SigChar V

Linking Relativistic Dark Matter And Dark Energy

If the mass/energy of the Universe is comprised of about 4% ordinary matter, 23% dark matter, and 73% dark energy, it probably would be logical to expect a link between the dark matter and the unknown dark energy that fuels an accelerating expansion of the Universe. One clue to support this theory is the very large percentage of the dark matter mass in the Universe that appears to be relativistic mass. See Chapter17 (SigChar N) and Chapter 24 (SigChar U) for an example of a possible coupling between dark matter and dark energy that may be the cause of an accelerating expansion of the Universe.

CHAPTER 26

SigChar W

How The First-Generation Stars May Have Been Ignited Without Dust Or Molecular Hydrogen

The European Southern Observatory (ESO) published an article about starburst galaxies on November 18, 2004, entitled, "Stellar Clusters Forming in the Blue Dwarf Galaxy NGC 5253." The researchers' astro-ph/0411486 paper is entitled, "The Star Cluster population of NGC 5253."[26] The following two excerpts from the ESO article provide some insight into the level of challenge presented to someone seeking an explanation for primordial star formation:

> Star formation begins with the collapse of the densest parts of interstellar clouds, regions that are characterized by comparatively high concentration of molecular gas and dust like the Orion complex and the Galactic Centre region. Since this gas and dust are products of earlier star formation, there must have been an early epoch when they did not yet exist.

The next paragraph continues:

> But how did the first stars then form? Indeed to describe and explain 'primordial star formation' without molecular gas and dust is a major challenge in modern astrophysics.

There are several possible explanations for primordial star formation (formation and ignition of the first stars) utilizing only atomic hydrogen and the relativistic proton dark matter theory/cosmology. Let us begin with the simplest.

Molecular hydrogen is needed for star formation. It is known that a mixture of 50% hydrogen ions and 50% neutral hydrogen atoms will form molecular hydrogen much faster than would neutral hydrogen atoms alone. Therefore, the bombardment/ionization of a galaxy's atomic hydrogen gas by the relativistic protons from the dark matter halo should facilitate the formation of hydrogen molecules on the surface of the enclosed galaxy, provided that no more than, say, 70% of the atomic hydrogen is ionized. The ideal ratio probably would be 50% hydrogen ions and 50% atoms to maximize the number of hydrogen ion-atom pairs to merge into molecules.

Another possible explanation for primordial star ignition utilizing only atomic hydrogen and the relativistic proton dark matter theory/cosmology involves four steps:

1. The relativistic dark matter protons and associated helium nuclei colliding with compressed interstellar clouds of hydrogen and helium atoms would generate muons that could create muonic atoms of hydrogen and of helium.

2. The muon orbits around the protons and helium nuclei are very small because the muons weigh 207 times as much as electrons. Therefore, the positive coulomb charges of the nuclei are well shielded by the closely orbiting negative muons, making the muonic atoms have a very low effective coulomb charge and thereby enabling them to collide with one another.

3. The colliding muonic atoms of hydrogen and helium could form muonic molecular ions with either two protons or one proton and one helium nucleus. These two types of ions are being orbited by only one muon since the net coulomb charge is so low that a second muon would not attach. By this means, muonic molecular ions are formed.

4. Subsequent bombardment of the muonic molecular ions by relativistic dark matter protons with energies of about 10^{15} eV and by high-velocity helium nuclei should be capable of triggering hydrogen fusion reactions and the ignition of new stars from the muonic molecular ions.

The author's star ignition and hydrogen fusion theory is based upon astronomical data that each 10^{15} eV cosmic ray proton striking the Earth's atmosphere produces perhaps hundreds to one thousand muons. See Chapter 27 (SigChar

X). (The cosmic ray protons actually produce pions which rapidly decay into muons which, in turn, decay less rapidly into electrons, etc., in a number of microseconds.)

For a number of decades, muons have been known to catalyze hydrogen fusion reactions by forming muonic molecular ions comprised of a proton plus a helium nucleus or deuterium or another proton orbited by a muon.[27,28] Muons are also known to catalyze multiple fusion reactions since they are not destroyed in the nuclear fusion process. A Google search for "muonic hydrogen fusion" leads to dozens of website references.

Star-related hydrogen fusion might have been feasible in the early Universe since relativistic dark matter protons were probably multitudinous and had energies much more than a thousand times higher than what can be achieved with man-made accelerators today. Also, catalytic muons were being produced in enormous quantities, and hydrogen/helium muonic molecular ions evolved as collision targets for the dark matter relativistic protons and helium nuclei spiraling into a galaxy from its dark matter halo.

An alternative explanation for primordial star formation is based upon the theory behind the starburst galaxy phenomenon, which is described in Chapter 46. It is known that the vast majority of starburst galaxies involve merging spiral galaxy clusters. According to the relativistic proton dark matter theory/cosmology, merging spiral galaxy clusters would subject the hydrogen in the galaxy disks to bombardment by protons from the galaxy clusters' DM halos with energies more than 10 times higher than protons from the bombarded galaxy's own DM halo. Such a higher energy proton bombardment should create more muons, muonic atoms, and muonic ions and would subject the latter to higher energy collisions, thereby facilitating hydrogen fusion and ignition of first-generation stars.

Three different methods of creating the first-generation stars have been presented. Depending upon conditions throughout the early Universe, any one of them could be favored in some space and at some time and possibly all three methods were involved in the early epoch.

The relativistic protons in dark matter halos have sufficient energy to trigger hydrogen fusion in galaxies. The website of the Princeton Plasma Physics Laboratory-Tokomak Fusion Test Reactor (http://www.pppl.gov/projects/pages/tftr.html) reports that by 1997 the Tokamak Fusion Test Reactor [for fusion of hydrogen isotopes] achieved a world record "plasma temperature of 510 million degrees centigrade -- the highest ever produced in a laboratory, and well beyond the 100 million degrees required for commercial [hydrogen] fusion." Note that in a 510 million degree plasma, the average kinetic energy of the plasma particles would be far below the kinetic energy of many galaxy-orbiting relativistic protons in the dark matter halo around a spiral galaxy.

The catalytic capability of muons in hydrogen fusion nuclear reactions has been known for about fifty years. In 1956, at the Los Alamos Meson Physics Facility and at U.C. Berkeley, Luis W. Alvarez and H. Bradner discovered the hydrogen-fusion-catalytic capability of the mu-meson, now called the muon, with the help of Edward Teller. They discovered that incoming muons were able to catalyze nuclear fusion between a proton and a deuterium nucleus (one proton and one neutron). Apparently, the muons were aiding the two types of nuclei to come close enough together for quantum tunneling to allow them to fuse, *even at room temperature*.

This nuclear fusion process was never commercialized because the proton bombardment energy required to produce the necessary muons was so great and the helium produced in the reaction captured so many muons (a principal source of energy loss) that the nuclear fusion process was very inefficient and impractical. However, this same or related process may be practical for creating hydrogen fusion in stars because of the extremely high energies of the multitudinous cosmic ray protons available in the Universe to bombard the muonic ions to trigger fusion and also to generate up to one thousand muons per bombarding proton, thereby overcoming and negating the muon-absorption-by-helium problem.

See SigChar J, K, and also see SigChar X, which discusses the role of the muons. In addition, see SigChar O regarding collisions of the relativistic dark matter protons with dust, photons, and hydrogen.

The thee above-described star formation methods cannot be explained by the generally accepted mainstream theory of star formation, where clouds of hydrogen molecules collapse anywhere in a galaxy under their own weight and are heated through compression to hydrogen fusion temperatures.

Relativistic Proton Dark Matter May Be The Source Of The Stars, Planets, And DNA Changes:

Relativistic proton dark matter may be the source of sunlight, starlight, the stars, planets, and DNA changes throughout the Universe. In this chapter, we have learned that proton dark matter probably feeds hydrogen fuel to all the galaxies and that it ionizes atomic hydrogen, thereby forming hydrogen molecules faster. The proton dark matter also creates muons that catalyze hydrogen fusion nuclear reactions, and it bombards and triggers hydrogen-based muonic ions that ignite the stars. Through these four functions, the relativistic dark matter protons seem to have created all the stars and, therefore, all the planets in the Universe.

CHAPTER 27

SigChar X

How The Later Generations Of Stars May Have Been Ignited Utilizing Both Dust And Molecular Hydrogen

In Chapter 26, SigChar W, note in the ESO article that the star formation regions "are characterized by comparatively high concentration of molecular gas and dust."[26] The question is, how are the hydrogen molecules and dust utilized to facilitate star formation?

It has been known for decades that when high energy cosmic ray protons strike the Earth's atmosphere, they generate muons. The Stanford SLAC website reports that a count of muons arriving at the Earth's surface from one cosmic ray proton collision totaled about 1,000 muons. For the purposes of this book, it is estimated that a 10^{15} eV proton collision in the Earth's atmosphere would generate hundreds of muons and as many as one thousand muons.

It is well known that negative muons have the same negative charge as the electron, weigh about 207 times as much, and can form muon-orbiting protons as atoms or as ions. It is also known that a muon can orbit one proton, forming an atom, or it can orbit a proton-proton pair or a proton-helium nucleus pair, forming a molecular ion with the muon closely orbiting the pairs because of its high mass, thereby, for example, pushing the proton and helium nucleus closer together or pushing two protons closer together. It is believed that if one of these two protons were replaced by deuterium or helium, hydrogen fusion could result, generating enormous amounts of energy, but the muon would be ejected unscathed to be able to catalyze additional similar fusion reactions until the muon decays into an electron.[27,28]

With the large flux of muons and electrons created by relativistic dark matter protons and helium nuclei bombarding the dust particles, helium, and hydrogen, and with the large flux of dark matter relativistic protons and helium nuclei passing through the region, a variety of particle collisions and particle configurations are feasible in this high-velocity muon-electron-proton plasma to trigger hydrogen fusion. For example, high-velocity, high-collision-cross-section helium nuclei being orbited by one muon, possibly could collide with a pair of protons being orbited by one muon, thereby triggering hydrogen fusion. The success of the fusion triggering action would depend upon how well the negative muons shield the positive coulomb charges of the nuclei involved so as to reduce the opposing coulomb forces sufficiently to permit high collision velocities of the charged-shielded nuclei.

Another hydrogen fusion mechanism, mentioned in the previous chapter, involves collisions with galactic hydrogen by relativistic protons and helium nuclei departing from the dark matter halo, so as to form muonic molecular ions. This step would be followed by the relativistic DM protons and high-velocity DM helium nuclei bombarding the newly formed muonic molecular ions containing either two protons or one proton and one helium nucleus, so as to trigger hydrogen fusion.

TABLE 1

RECAP

Galaxy-Orbiting Relativistic Protons Are Expected To Have Signature Characteristics (SigChar) Related To The Following Subjects:

PART I – Signature Characteristics A - X

SigChar A	Dark Matter (DM) Proton Energies
SigChar B	The Milky Way's Magnetic Fields
SigChar C	Larmor Radius Equation
SigChar D	The Milky Way's Dark Matter Halos And Proton Energies
SigChar E	Paths Of Protons
SigChar F	Proton Streams Creating Magnetic Fields
SigChar G	Proton Flux And Kinetic Energy In Halos
SigChar H	Proton Relativistic Mass Losses From Synchrotron Radiation
SigChar I	Magnetic Bulges Leading To Increased Synchrotron Radiation From Protons
SigChar J	Why Dark Matter Halo Protons Enter Their Enclosed Galaxy And Lose Relativistic Mass
SigChar K	Protons/Helium Nuclei Collisions With Hydrogen Clouds
SigChar L	Linearly Rising Rotation Curves Indicating LSB Dwarf Galaxy DM Halos Are "Weakly Centrally Concentrated" (i.e., "Hollow")

TABLE 1

(Continued)

SigChar M	An Explanation For The Two "Knees" And "Ankle" Of The Cosmic Ray Energy Spectrum
SigChar N	Proton Synchrotron Radiation Losses And Proton Collision Losses Possibly Could Lead To An Accelerating Expansion Of The Universe
SigChar O	Radiating DM Halo Protons Become Cosmic Ray Protons
SigChar P	Long, Large, DM Filaments Creating Galaxy Clusters
SigChar Q	Mature Galaxies In A Young Universe
SigChar R	Conservation Of Angular Momentum
SigChar S	No DM Cusps In Nuclei Of Galaxies
SigChar T	Explanations For LSB Dwarf Galaxies' Low Star Formation Rates (SFRs) And For Massive Galaxies' Very High SFRs
SigChar U	The Relativistic Energy Of All The Protons In The Universe May Provide The Energy For The Accelerating Expansion Of The Universe
SigChar V	Linking Relativistic Dark Matter And Dark Energy
SigChar W	How The First-Generation Stars May Have Been Ignited Without Dust Or Molecular Hydrogen
SigChar X	How The Later Generations Of New Stars May Have Been Ignited Utilizing Both Dust And Molecular Hydrogen

CHAPTER 28

Tentative Conclusions, Insights, Explanations, And Possible Astrophysical Discoveries

A new research process designated *dark matter relationism* has been developed by the author based upon finding astronomically based Cosmic DM Mysteries of the Universe that may be created or influenced by or have a special relationship with possible dark matter candidates. The Cosmic DM Mysteries represent various mysterious or unexplainable astrophysical or cosmological phenomena.

As stated earlier, *dark matter relationism* is epitomized in the author's use of cosmic mysteries and *relationism* to tentatively identify dark matter and then to confirm its validity by using this same dark matter candidate to provide plausible explanations for more cosmic mysteries, including some previously *not known to be related to dark matter*.

As described in the following chapters, a list of 14 relevant and plausible Cosmic DM Mysteries of the Universe was developed by the author toward the end of 2004. This list was then used to establish a number of constraints regarding the nature and characteristics of the long-sought dark matter particles. A dark matter candidate was then found that best conformed to the constraints established and imposed by the 14 Cosmic DM Mysteries.

The author will show in the following chapters that this same conforming dark matter candidate provides plausible explanations for and discloses possible discoveries related to a number of astrophysical and cosmological phenomena, including the following:

1. The accelerating expansion of the Universe.
2. The "knees" and "ankle" of the cosmic ray energy spectrum graph.
3. The low star formation rates of LSB dwarf galaxies; the medium-level SFRs of the Milky Way; and the very high SFRs of large, massive galaxies.
4. The ignition of hydrogen fusion reactions in the first generation stars.
5. The source of magnetic fields in spiral galaxies.
6. How dust particles facilitate hydrogen fusion in stars.
7. The four locations within the Local Group and the Virgo Supercluster from which Earthbound cosmic ray protons depart.

With the conforming dark matter candidate tentatively identified as *galaxy-orbiting relativistic proton* dark matter, in the following chapters, this candidate and its two dozen Signature Characteristics will be utilized and applied to the 14 Cosmic DM Mysteries to provide plausible scientific explanations for each them, including the seven astrophysical and cosmological phenomena listed above.

The Nature And Characteristics of Galaxy-Orbiting Relativistic Proton Dark Matter:

- The number of the strongly interacting relativistic dark matter protons orbiting the Milky Way probably totals about 11 orders of magnitude greater than the number of cosmic ray protons entering the Milky Way annually from its dark matter halo. This is based upon an estimate that during the past 10 billion years, perhaps about 10% of the relativistic proton dark matter has been transformed into ordinary hydrogen under a top-down theory of galaxy formation.

- It is generally accepted by cosmologists that ordinary matter in the Universe totals about 4% of the total mass/energy and dark matter totals about 23% of the total mass/energy and, therefore, the mass energy of dark matter in the Universe is about six times larger than that of ordinary matter. This same ratio appears to be plausible within the relativistic proton dark matter theory/cosmology.

- If the total number of relativistic dark matter protons in the Universe is not more than 15% of the total number of baryons in the Universe, the relativistic proton dark matter theory would be compatible with the proton limitations of cosmologists. Further, if the average relativistic mass of the dark matter protons is about 50 times the rest mass of a proton and the number of protons is, say, about 12% of the number of baryons in the Universe, the dark matter mass would total about six times that of the non-relativistic ordinary matter baryon mass in the Universe. This matches the cosmologists' estimates mentioned above. In order to achieve 50 times the rest mass of a proton, the average proton energy required would be about 5×10^{10} eV, which is at the low end of the cosmic ray proton energy range of about 1×10^{10} eV to 5×10^{19} eV.

- Relativistic proton motion follows Coulomb's law and the Larmor Radius equation (see Chapter 6). The Larmor Radius of the relativistic proton paths is directly proportional to the proton energy and inversely proportional to the orthogonal magnetic field strength.

- Relativistic protons create muons and electrons through collisions with dust, hydrogen, helium, and CMB photons. The protons then can combine with these electrons to form hydrogen.

- Some relativistic dark matter protons are transformed over millions to billions of years into hydrogen, and others become cosmic ray protons plunging into galaxies.

- It is generally accepted that intersecting, long, large, relativistic proton dark matter filaments that crisscross the cosmos create galaxy clusters at their intersections.

This process has the characteristics of a top-down theory of galaxy formation.

- The relativistic dark matter protons' spiral paths in the dark matter halo are determined by the Larmor Radius equation, shown in Chapter 6.

- Relativistic proton dark matter halo morphology is determined by the Larmor Radius equation (and coulomb forces); i.e., they would be spherical/ellipsoidal dark matter halos with "hollow" cores. The "hollow" core of the dark matter halo of the Milky Way should be expected because orbiting dark matter protons with energies much less than 3×10^{15} eV would tend to lose kinetic energy through synchrotron radiation at above-average rates, leave the core region of the dark matter halo, and enter the Milky Way as cosmic ray protons or as hydrogen.

- Relativistic protons in Larmor orbits increase their synchrotron radiation loss when they move into a higher magnetic field or if they slow down through collisions or through synchrotron energy losses.

- Under the top-down theory of galaxy formation, proto-galaxies form and grow through the accretion of hydrogen converted from relativistic protons. The galaxy disk enclosed within a dark matter halo may be larger or smaller than the relativistic proton dark matter halo's "hollow" core diameter. If the galaxy disk is smaller, the galaxy should exhibit a low SFR; and if the disk is much larger, such a massive galaxy should exhibit a very high SFR. The decline of SFRs over time would be very small for small galaxies because their dark matter halos could provide adequate replenishment hydrogen to compensate for the hydrogen consumption of its stars.

However, for large massive galaxies with their very high SFRs, the replenishment hydrogen from their dark matter halos could be inadequate after a time period, subjecting their SFRs to declines that are large and rapid.

- Relativistic protons in Larmor orbits create galactic magnetic fields through the astrophysical dynamo effect. That is, the proton paths and magnetic fields are doubly interrelated, in that each is determined by the other.

- Relativistic protons in Larmor orbits emit synchrotron radiation continuously and, thereby, their orbital velocities and relativistic mass decline continuously. Thus, the dark matter halo mass declines continuously and the separation velocity between galaxy clusters should increase under the Law of Conservation of Linear Momentum, to yield an accelerating expansion of the Universe.

- Strongly interacting and randomly moving multitudinous relativistic protons may evolve into relativistic proton dark matter galaxy halos and galaxy cluster halos over millions to billions of years through *emergent evolution*[1] principles (or emergence or collective self-organization).

---

[1] *Emergent evolution* (also known as emergence, astrophysical emergence, or collective self-organization) is a theory that new characteristics and qualities appear in the evolutionary process at more complex organizational levels that cannot be predicted by studying less complex levels of organization, but are determined by a rearrangement of pre-existing entities. Also, for emergent evolution principles to succeed, the pre-existing entities must be strongly interacting in conjunction with collective self-organization. Relativistic protons are strongly interacting, but cold dark matter WIMPs are not. See Chapter 44.

Of significance, the relativistic proton dark matter theory and cosmology appear to be compatible with (1) astronomers' estimates for the size of the Milky Way and its dark matter halo, (2) the strength of the extragalactic magnetic field, (3) penetration of the DM halo into the Milky Way, and (4) the cosmic ray proton energy distribution at the Earth.

In Part II of this book, each of the next 14 chapters begins with a description of one of the 14 Cosmic DM Mysteries, followed by the alphabetic identification of the Signature Characteristics that are relevant to the conforming dark matter candidate. Eleven subsequent chapters in Part III of this book deal with an additional 11 Cosmic DM Mysteries that came to the author's attention subsequent to April 2005.

Additional References:

For additional information related to cosmic rays and ultra-high energy cosmic rays, see references *18, 28, 29, 30*, and *31*, which include astrophysics papers by A. A. Watson and J. W. Cronin. For an astrophysics paper summarizing the status of dark matter research as of early 2004 by Sir Martin Rees, see reference *32*.

Comprehending And Decoding The Cosmos

PART II

CHAPTER 29

Cosmic DM Mystery #1
Spiral Disk Galaxies Have Spherical Dark Matter Halos

Relativistic Proton Dark Matter Particles Could Form Spherical DM Halos Around Spiral Galaxies And DM Halos Around Galaxy Clusters

See SigChar C, D, E, G, L, N, and Chapter 44.

The author believes that the dark matter halos around spiral galaxies have roughly defined outer diameters and "hollow" core diameters determined by the galactic and extragalactic magnetic field magnitudes and the energy spectrum of the relativistic protons. The kinetic energies of the protons orbiting spiral galaxy *clusters* are probably about 30 times higher than those orbiting spiral *galaxies*, as determined by the Larmor Radius equation, the cosmic ray proton energy distribution at the Earth, the magnetic field strength, and the size of a galaxy cluster compared to the size of a spiral galaxy disk. (See Chapter 50.)

Also, the author believes that the outer diameter and core diameter size of dark matter halos are not significantly affected by the amount of galaxy mass enclosed, since electromagnetic forces are tens of orders of magnitude greater than gravitational tidal forces.

Astronomical data indicate that dark matter forms enormous spherical halos around spiral galaxies, which extend outward from the galaxy to 10 to 20 times the radius of the galaxy disk. The morphology of the spherical DM halos could be affected by the protons' coulomb forces.

The author believes that dark matter halos around spiral galaxies and around their galaxy clusters are created by a phenomenon the author calls *astrophysical emergence*, also referred to as *emergent evolution*. The footnote paragraph near the end of Chapter 28 briefly describes this phenomenon, and Chapter 44 is entirely devoted to *astrophysical emergence*. The author is not aware of any literature references that apply the principle of *emergence* to the formation of galaxies or dark matter halos.

CHAPTER 30

Cosmic DM Mystery #2
Accelerating Expansion Via Conserving DM Momentum

Relativistic Proton Dark Matter Particles Could Cause An Accelerating Expansion Of The Universe And Possibly Store Dark Energy

See SigChar C, H, I, N, U, and V.

An accelerating expansion of the Universe may come about as a result of kinetic energy loss of the relativistic dark matter protons orbiting spiral galaxy clusters and orbiting the clusters' spiral galaxies as a result of synchrotron radiation losses and kinetic energy losses from collisions with photons, dust, hydrogen, and helium.

This leads to a reduced relativistic mass of the galaxy clusters that are receding from each other according to the Hubble Law. Their declining mass should cause them to accelerate to maintain their linear momentum under the Law of Conservation of Linear Momentum.

Also, it is possible that the total kinetic energy and/or the energy equivalent of the relativistic mass of all the dark matter relativistic protons in the Universe could be linked to the mysterious "dark energy" since a portion of that same energy seems to be consumed in conjunction with an accelerating expansion of the Universe.

CHAPTER 31

Cosmic DM Mystery #3
Hydrogen Derived From DM Cosmic Ray Protons

Relativistic Proton Dark Matter Particles Could Be Transformed Into Low-Velocity Hydrogen, Protons, Or Proton Cosmic Rays

See SigChar O, W, and X.

Low-velocity hydrogen and protons and relativistic cosmic ray protons are all prevalent in galaxies. The most logical source of all three of these types of this ordinary matter would be relativistic dark matter protons that had lost kinetic energy through synchrotron radiation and/or collisions with dust, photons, helium, or hydrogen.

When relativistic protons in a dark matter halo give up some kinetic energy through synchrotron radiation or through collisions with dust, photons, or molecules or atoms of hydrogen or helium, they decelerate into slower moving relativistic protons. When the kinetic energies of some of the relativistic protons in the Milky Way's dark matter halo are reduced much below 3×10^{15} eV, their synchrotron radiation losses accelerate and their kinetic energy, relativistic mass, and radius of curvature of their spiral orbits decline. The reduced-kinetic-energy relativistic dark matter protons then depart the dark matter halo and move into the enclosed galaxy and eventually plunge into star systems as hydrogen or as cosmic ray protons.[12]

Let us see how hydrogen could be created. Relativistic dark matter protons and helium nuclei colliding with compressed interstellar clouds of hydrogen and helium atoms should generate muons. It is estimated that each 10^{15} eV cosmic ray proton striking the Earth's atmosphere produces perhaps hundreds to one thousand muons. They actually produce pions that rapidly decay into muons that, in turn, decay less rapidly into electrons in a number of microseconds. The electrons are thus produced in large numbers per relativistic proton and, therefore, should be available in quantity to combine with the decelerating protons to form hydrogen.

CHAPTER 32

Cosmic DM Mystery #4
Magnetic Fields Derived From DM Cosmic Ray Protons

Relativistic Proton Dark Matter Particles Could Create The Magnetic Fields Within And Around Spiral Galaxies

See SigChar F and U.

The creation of magnetic fields surrounding spiral galaxies requires the flow of coulomb charges through space. Relativistic dark matter protons orbiting galaxies will create such magnetic fields through the astrophysical dynamo effect. The basic principle of the astrophysical dynamo effect is that relativistic protons in Larmor orbits create the magnetic fields. These same magnetic fields, in turn, determine the proton paths during an *emergent evolution* period involving millions to a few billion years. At the end of this evolution period, mutually compatible steady-state solutions will have been established for both the morphology and magnitude of the magnetic fields and the paths of the orbiting proton streams.

Electrons do not play a significant role in establishing magnetic fields in space because their synchrotron radiation losses are 11 trillion times greater than that for protons, and their resultant rapid energy/velocity decline precludes them from being a significant contributor to the galactic and extragalactic (intergalactic) magnetic fields.

CHAPTER 33

Cosmic DM Mystery #5
Intersecting DM Filaments Create Galaxy Clusters

Relativistic Proton Dark Matter Particles Could Be Concentrated In The Long, Large Filaments Of Dark Matter (Announced By NASA 9/9/04) That Form Galaxy Clusters Where The DM Filaments Intersect

See SigChar P, W, and X.

The September 9, 2004 news release from NASA (and Harvard) entitled, "Motions in nearby galaxy cluster reveal presence of hidden superstructure,"[33] regarding Chandra x-ray images of the Fornax cluster, states:

> Astronomers think that most of the matter in the universe is concentrated in long large filaments of dark matter and that galaxy clusters are formed where these filaments intersect.

The researchers' related paper astro-ph/0406216 is entitled, "The Chandra Fornax Survey - I: The Cluster Environment."[34] This astronomically established filamentary description of dark matter appears to be much more compatible with the relativistic proton dark matter theory than the cold dark matter theory. It seems highly unlikely that the DM filamentary structure could be created by very slow moving, weakly interacting (only through gravitational tidal forces) particles.

The vision of DM filaments crisscrossing the cosmos gives the impression of high-velocity particles, while the crashing of intersecting DM filaments creating galaxy clusters gives the impression of a top-down theory of galaxy formation. Both of these impressions point toward and lend support to the relativistic dark matter theory/cosmology.

Furthermore, the theoretical WIMPs, being non-baryonic, cannot be transformed into hydrogen and helium where the DM filaments intersect, whereas the relativistic protons and helium nuclei, being baryonic, can provide hydrogen and helium to the galaxy clusters where the filaments intersect.

For the above reasons, the September 2004 reports of the DM filaments seemed to be very supportive of Drexler's DM theory and encouraged him to write this book sequel.

CHAPTER 34

Cosmic DM Mystery #6
Mature Galaxies Discovered In The Very Early Universe

Relativistic Proton Dark Matter Particles Could Create Large, Mature, Spiral Galaxies Less Than 2.5 Billion Years After The Big Bang

See SigChar Q.

Relativistic protons could create large, mature, spiral galaxies less than 2.5 billion years following the Big Bang if galaxy clusters are created via the top-down theory of crashing, intersecting, dark matter filaments followed by galaxy formation from the dark matter remnants, as indicated in the previous chapter. Then, galaxies would form and grow through the accretion of hydrogen and protons from the relativistic proton dark matter remnants, onto proto-galaxies.

The recent discovery of the existence of mature galaxies only about 2.5 billion years after the Big Bang[20,21,22] (and confirmed by the Carnegie Observatories on March 10, 2005)[35,36] can be explained using the relativistic proton dark matter theory/cosmology. This top-down theory of galaxy formation involves relativistic protons that create and combine with electrons to form the galactic hydrogen, but it raises questions about the cold dark matter bottom-up theory of galaxy formation, which involves only very slow-moving, weakly interacting, non-baryonic particles.

The above references *20* and *21* are articles in the July 2004 issue of Nature, entitled, "A high abundance of massive galaxies 3 - 6 billion years after the Big Bang" and "Old galaxies in the young Universe." The Carnegie Observatories had announced in a news release on March 10, 2005[35,36] that:

> Astronomers have found distant red galaxies -- very massive and old -- in the universe when it was only 2.5 billion years post Big Bang.

This news would imply that these old red galaxies were formed very much earlier as new blue galaxies.

Also supporting the relativistic proton dark matter theory is the discovery, reported in Chapter 51, of 800 young galaxies located 12.5 billion light years away, that were born about 1.2 billion years after the Big Bang.

CHAPTER 35

Cosmic DM Mystery #7
Dark Matter Spherical Cored Halos Have "Hollow" Cores

Relativistic Proton Dark Matter Particles Could Create Spherical DM Halos Having Predictable Outer And "Hollow" Core Diameters

See SigChar A, B, C, D, E, F, G, and Chapter 44.

The author believes that orbiting relativistic protons in the extragalactic magnetic field, during a period of millions to billions of years, could create spherical dark matter halos around spiral galaxies through the principle of *astrophysical emergence*. (See Chapter 44.) These spherical halos would have roughly predictable outer and "hollow" core diameters determined by the kinetic energies of the relativistic protons and the galactic and extragalactic magnetic field strengths, through use of the Larmor Radius equation.

Relativistic protons have strongly self-interacting characteristics that play an important and necessary role in *astrophysical emergence*. If one is trying to explain the creation of dark matter halos through the concept of *astrophysical emergence* utilizing either relativistic protons or cold dark matter WIMPs as the simple microscopic entities, relativistic protons with their strong and multiple self-interacting forces (coulomb, electromagnetic, and nuclear) would be a far more promising candidate than the CDM theory's theoretical *weakly interacting* massive particles.

CHAPTER 36

Cosmic DM Mystery #8
Source of Spiral Galaxies'/Halos' Angular Momentum

Relativistic Proton Dark Matter Particles Could Provide Angular Momentum To Spiral Galaxies And Their DM Halos

See SigChar C and R.

Galaxy-orbiting relativistic protons creating magnetic fields via the astrophysical dynamo effect eventually will achieve a steady-state, dynamical configuration with significant angular momentum, which can be transferred to a spiral galaxy and its dark matter halo under the Law of Conservation of Angular Momentum. Spiral galaxies embedded within or overlapping the "hollow" cores of dark matter halos (essentially orbiting proton streams) represent one such steady-state configuration with a high angular momentum.

The cold dark matter theory does not offer any explanation as to the role or roles of slow-moving, non-baryonic, weakly interacting WIMPs in providing angular momentum to either the spiral galaxies or their dark matter halos. This could imply a lack of a link or coupling between WIMPs and DM halos.

CHAPTER 37

Cosmic DM Mystery # 9
No Central Dark Matter Cusp Found In Spiral Galaxies

Relativistic Proton Dark Matter Particles Could Create Galaxies Without A Central DM Density Cusp

See SigChar B, C, E, and S.

Orbiting dark matter relativistic protons under the influence of a magnetic field generally will be constrained in orbits and will not move toward the nucleus of a galaxy unless they have lost a large percentage of their kinetic energy. In this case, the protons probably would interact magnetically and collide with one of the multitudinous (200 billion stars for the Milky Way) star systems they would encounter on their way toward the galaxy nucleus. Also, the coulomb forces of protons would prevent them from forming a cusp of protons at the galaxy nucleus.

The Sun, a typical star, has a magnetic field strength roughly on the order of magnitude of about one gauss, and that of the earth is about 0.5 gauss. Thus, the interplanetary magnetic field strength would be roughly of the order of 0.5 gauss. The synchrotron radiation losses for a relativistic cosmic ray proton passing near a star could be of the order of 200,000 times greater than that for interstellar space. Also, a star's magnetic field would tend to trap a proton passing nearby.

Therefore, the dark matter relativistic protons would not normally form a dark matter mass density cusp at the nucleus of a galaxy.

No dark matter mass density cusps are found in the centers of spiral galaxies. However, the cold dark matter theory predicts the formation of DM mass density cusps at the centers of spiral galaxies. The lack of DM mass density cusps has raised questions about the validity of the CDM theory.

CHAPTER 38

Cosmic DM Mystery #10
LSB Dwarf Galaxies Have Low Star Formation Rates

Relativistic Proton Dark Matter Particles Could Create A Starless Galaxy Or An LSB Dwarf Galaxy With Low Star Formation Rates

See SigChar A, B, C, D, E, G, J, L, M, and T.

The author believes that the diameter of a galaxy disk embedded within the "hollow" core of a dark matter halo can be larger, smaller, or the same size as the inner diameter of the "hollow" core. Based upon the Milky Way, it is estimated that the galactic magnetic field strength of spiral galaxies is perhaps in the range of 2,000 times higher than the extragalactic magnetic field surrounding the dark matter halo of a spiral galaxy.

The low star formation rates for starless dark galaxies and LSB dwarf galaxies can be explained if the galaxy-orbiting relativistic protons in the dark matter halo are normally utilized either to ignite the hydrogen fusion nuclear reaction of new stars or to provide proton or hydrogen fuel to galaxies for its stars. It would follow that the low star formation rate could occur if the galaxy disk diameter is smaller than the DM halo "hollow" core diameter and, therefore, the relativistic protons in the dark matter halo near the "hollow" core would not be subjected to the full magnitude of the much higher magnetic field of the enclosed galaxy. In this case, the resultant synchrotron radiation losses of the relativistic dark matter halo protons would be smaller than normal, thereby reducing the number of 1×10^{16} eV protons that would move from the dark matter halo into the galaxy to facilitate the formation of stars.

The "hollow" core of the dark matter halo of the Milky Way was created at an earlier time, when the Milky Way's disk was much smaller than it is today. At that earlier time, orbiting DM protons with energies much less than 3×10^{15} eV would lose a significant portion of their kinetic energy through synchrotron radiation energy loses and leave the core region of the dark matter halo, thereby enlarging the diameter of the "hollow" core.

CHAPTER 39

Cosmic DM Mystery #11
LSB Galaxies Have Inclining Star Rotation Curves

Relativistic Proton DM Particles Could Lead To Linearly Rising Rotation Curves For LSB Dwarf Galaxies And To Flat Rotation Curves For Spiral Galaxies

See SigChar L and consider Chapter 15 as being integral with this chapter.

The recently announced (February 11, 2005)[17] linearly rising rotation curves of LSB dwarf galaxies, compared to the flat rotation curves for large spiral galaxies, indicated to researchers that the dark matter halos of LSB dwarf galaxies are "weakly centrally concentrated."

This supports the author's previously developed concept of the "hollow" cores of dark matter halos and their core-size relationship to the size of the enclosed galaxy disks. One might conclude that if the enclosed galaxy disk is smaller than the "hollow" core diameter of the halo, the star formation rate probably will be low. If the galaxy disk diameter overlaps the "hollow" core of the dark matter halo, the galaxy probably could have a high star formation rate. If the galaxy overlap is even greater, such as for a large massive galaxy, the SFR could be very high for a period of time. If a galaxy disk is smaller than a typical LSB galaxy disk or if its galactic magnetic field is unusually low, it could appear to be a "dark galaxy."

A dark dwarf galaxy can remain a "dark galaxy" accreting hydrogen from its dark matter halo onto the enclosed small galaxy disk for millions to billions of years, until the galaxy disk grows sufficiently in size to overlap the "hollow" core of its dark matter halo and star production could begin. The author believes that diminutive galaxy I Zwicky 18[24] (3,000 light-years across) is of this type since its stars formed only about 500 million years ago, yet it is surrounded by mature galaxies billions of years old. It is about 45 million light-years away from the Sun.

There are well over 100 astro-ph papers on the physics arXiv on the subject of *cored* dark matter halos/haloes, the formation of which is not explainable by the CDM theory, but is easily explainable by the relativistic proton dark matter theory/cosmology.

The well-known flat galaxy rotation curves discovered by Vera Rubin, involving the stars and hydrogen in spiral galaxies, indicate that the spherical mass of dark matter contained within a sphere of radius r from the nucleus of the spiral galaxy, increases linearly with radius r through the galaxy and into the halo.[3,4] This means that the dark matter mass density at radius r must be declining approximately as the square of the radius r. This astronomical feature of dark matter halo mass also seems to be inherent in a relativistic proton dark matter halo, as is explained in the following paragraphs.

Galaxy-orbiting relativistic proton streams in dark matter halos, as pointed out in Chapter 10 (SigChar G), are expected to have a rapidly declining proton flux density as a function of proton path radius. That is, the highest proton flux density would be near the "hollow" inner core of the dark matter halo, and the lowest proton flux density would be near the dark matter halo's outer diameter.

More specifically, while the *particle* flux density of the relativistic protons falls as a power law of their energies, with the exponential decline ranging between 2.7 and 3, as determined from the cosmic-ray energy distribution graph[5,18] (see Appendix B, Slide #17), the relativistic mass of each proton rises linearly with kinetic energy and is highest at the dark matter halo's outer diameter.[5]

This leads to the rough approximation that the *mass* flux density of the orbiting protons, within the galaxy and within its DM halo, falls approximately inversely as the square of the radius r from the nucleus of a spiral galaxy radially outward, thereby approximately satisfying Vera Rubin's flat-rotation-curves requirement.

CHAPTER 40

Cosmic DM Mystery #12
Galaxy Hydrogen Is Replenished From Halo Dark Matter

Relativistic Proton Dark Matter Particles Could Form About 80% To 85% Of The Mass Of The Universe, The Remainder Being Hydrogen, Helium, Etc.

See SigChar A, O, V, and the Introduction.

What kind of dark matter particles could form about 80% to 85% of the mass of the Universe, the remainder being hydrogen, helium, etc? The DM relativistic proton fits these astronomical/cosmological data constraints since it can play a significant role both with the 80% to 85% dark matter and with hydrogen and helium categories. It fits the hydrogen category easily since the relativistic protons and the helium nuclei are transformed into hydrogen and helium when they collide with dust, atoms, molecules, or photons, thereby producing transient muons and permanent electrons for the transformation of the protons into hydrogen and helium nuclei into helium.

For many years, baryons have been ruled out as a DM candidate because the primordial nucleosynthesis calculations and other cosmological considerations indicate a low *particle abundance* of baryons in the Universe. However, this argument does not address the large quantity of relativistic protons/baryons in the Universe (see Chapter 24). For example, a 15% ratio of relativistic protons to non-relativistic baryons satisfies the particle-abundance-maximum constraint and also satisfies the mass-abundance-minimum constraint because the DM particles, having a large relativistic mass, still enable them to form 80% to 85% of the mass of the Universe.

That is, they would meet this condition if the average DM particle's relativistic mass were only about 50 times the rest mass of a proton, represented by an average proton kinetic energy of about 5×10^{10} eV.

Note that cosmic ray proton energies range between 1×10^{10} eV to 5×10^{19} eV and that protons at the 5×10^{10} eV energy level are at the low energy end of the range where the cosmic ray proton flux is highest. (See Appendix B, Slide #17.)

CHAPTER 41

Cosmic DM Mystery #13
Dark Matter, Hydrogen, Helium, And Muons Create Stars

Relativistic Proton Dark Matter Particles Could Ignite Hydrogen Fusion Reactions Of First-Generation Stars Using Only Hydrogen And Helium Atoms, And Of Second-Generation Stars Using Hydrogen Molecules, Helium, And Dust As Well

See SigChar A, K, W, and X and consider Chapters 26 and 27 as being integral with this chapter.

Star Ignition Using Hydrogen Atoms:

It is known that dwarf galaxies having a large amount of atomic hydrogen and dark matter may exhibit a subnormal star formation rate. Also, many astrophysicists believe that "to explain primordial star formation without molecular [hydrogen] gas and dust is a major challenge in modern astrophysics."[26] See Chapter 26 (SigChar W) and also reference *26*.

Yet, astrophysicists know that primordial star formation did take place. The author believes that the relativistic proton dark matter theory/cosmology offers several possible explanations for the formation of the first stars, utilizing only atomic hydrogen. Let us begin with the simplest.

It is known that a mixture of 50% hydrogen ions and 50% neutral hydrogen atoms will form molecular hydrogen much faster than would neutral hydrogen atoms alone. Therefore, the bombardment/ionization of the atomic hydrogen gas in a spiral galaxy by the relativistic protons from the dark matter halo should facilitate the formation of hydrogen molecules in the galaxy, provided that no more than, say, 70% of the atomic hydrogen is ionized. The ideal ratio would be 50% hydrogen ions and 50% atoms to maximize the number of hydrogen ion-atom pairs to merge into molecules.

As another explanation under the relativistic proton dark matter theory/cosmology, primordial star ignition utilizing only atomic hydrogen might have occurred through the following four steps:

1. Galaxy-orbiting relativistic dark matter protons and associated helium nuclei colliding with compressed interstellar clouds of hydrogen atoms and helium atoms

would generate muons, which in turn would create muonic atoms of hydrogen and of helium with muons replacing the electrons.

2. Muons in muonic atoms orbit the protons and helium nuclei at much smaller radii than electrons because the muons weigh 207 times as much as an electron. Therefore, the positive coulomb charges of the nuclei are well shielded by the closely orbiting negative muons, making the muonic atoms have a very low effective charge and thereby enabling them to collide with each other.

3. The colliding muonic atoms of hydrogen and helium form muonic molecular ions with either two protons or one proton and one helium nucleus. These two types of muonic molecular ions are being orbited by only one muon since the net coulomb charge is so low that a second muon typically would not attach. By this means, muonic molecular ions are formed.

4. Subsequent bombardment of the muonic molecular ions by relativistic protons with energies of about 10^{15} eV and by the large relativistic helium nuclei should be capable of triggering hydrogen fusion reactions and the ignition of new stars from either one or both types of the muonic molecular ions. Also, high-speed collisions between the muonic molecular ions themselves could also participate in the hydrogen fusion nuclear reactions.

The author's star ignition and hydrogen fusion theory is based upon astronomical data that each 10^{15} eV cosmic ray proton striking the Earth's atmosphere produces hundreds to perhaps one thousand muons. See SigChar X. (The cosmic ray protons actually produce pions that rapidly decay into muons that, in turn, decay less rapidly into electrons, etc., in a number of microseconds.)

For a number of decades, muons have been known to catalyze hydrogen fusion reactions by forming muonic molecular ions comprised of a proton plus a helium nucleus or deuterium or another proton orbited by a muon.[27,28] Muons are also known to catalyze multiple fusion reactions because they are not destroyed in the nuclear fusion process. A Google search for "muonic hydrogen fusion" leads to a number of website references.

Star-related hydrogen fusion might have been feasible in the early Universe since relativistic dark matter protons were probably multitudinous and had energies much more than a thousand times higher than what can be achieved with man-made accelerators today. Also, catalytic muons were being produced in enormous quantities, and hydrogen and helium muonic molecular ions evolved as collision targets for the dark matter relativistic protons and high-velocity helium nuclei spiraling into a galaxy from its dark matter halo. See SigChar J, K, W, and X.

There is also the possibility that some primordial star formation can be attributed to the merging of early galaxy clusters, thereby forming starburst galaxies. See Chapter 46.

Star Ignition With Hydrogen Gas Molecules And Dust Would Generate Even More Muons:

The relativistic proton dark matter and accompanying relativistic helium nuclei could enter into collisions with the molecular and atomic hydrogen and helium gas and with dust particles in the enclosed galaxy. The higher collision cross sections of the dust and molecules, compared to those of the atoms, could lead to a significantly higher rate of muon formation, which in turn could facilitate and expedite star formation. That is, the larger volume of muons would replace more electrons in the hydrogen and helium atoms, in the hydrogen and helium molecules, and in the hydrogen/helium molecules. This would lead to the formation of more muonic ions with a muon orbiting two protons or a proton-helium nucleus pair, pushing them closer together and creating a collision target for the bombarding relativistic protons and high velocity helium nuclei arriving from the cored dark matter halo.

Subsequent bombardment of the larger volume of muonic molecular ions by relativistic protons with energies of about 10^{15} eV and by the high-velocity helium nuclei should be capable of triggering a higher rate of hydrogen fusion reactions and a higher rate of ignition of new stars from either one or both types of the muonic molecular ions. Also, high-speed collisions between the muonic molecular ions themselves also could participate in the hydrogen fusion reactions.[27]

As discussed in Chapter 26, the catalytic capability of muons in hydrogen fusion nuclear reactions has been known for about fifty years. In 1956, at the Los Alamos Meson Physics Facility and at U.C. Berkeley, Luis W. Alvarez and H. Bradner discovered the hydrogen-fusion-catalytic capability of the mu-meson, now called the muon, with the help of Edward Teller. They discovered that incoming muons were able to catalyze nuclear fusion between a proton and a deuterium nucleus (one proton and one neutron). Apparently, the muons were aiding the two types of nuclei to come close enough together for quantum tunneling to allow them to fuse, *even at room temperature.*

CHAPTER 42

Cosmic DM Mystery #14
Earthbound Cosmic Ray Protons Depart From 4 Locations

Relativistic Protons Create The First "Knee" At 3×10^{15} eV, The Second "Knee" Between 10^{17} eV And 10^{18} eV, And The Ankle At 3×10^{18} eV Of The Cosmic Ray Proton Energy Distribution Near Earth

See SigChar C, D, M, and Chapters 45, 47, and 50.

The author's relativistic proton dark matter theory posits that the highest energy dark matter protons are orbiting galaxy clusters and superclusters, a high energy relativistic proton group orbits galaxies, and a lowest energy relativistic proton group circulates within the enclosed galaxies. From the proton-flux versus proton-energy distribution measured near the Earth (see Appendix B, Slide #17) and the Larmor Radius equation, one can determine that the relativistic proton flux density is highest within the Milky Way, is second highest within the Milky Way's DM halo, is third highest in the DM halo of the Local Group of galaxies, and is lowest in DM halos of the Virgo Supercluster.

The author believes that the first "knee" of the above-referenced energy distribution graph at 3×10^{15} eV probably comes about because the protons that arrive at the Earth from the Milky Way have energies below the first "knee" energy level, and those that come from the Galaxy's dark matter halo have energies above the first "knee" energy level. The second "knee" at about 3×10^{17} eV probably separates protons arriving from the Galaxy's dark matter halo from those in the dark matter halo of the Local Group, which could have energies at or above 3×10^{17} eV. The "ankle" ultra-high energy is at or above 3×10^{18} eV.

The author believes that the four regions of the cosmic ray energy distribution graph (see Appendix B, Slide #17) have four different slopes because they represent relativistic protons arriving from four different locations, each with its own proton energy levels and proton flux densities: (1) the Milky Way, (2) its dark matter halo, (3) the dark matter halo of the Local Group of galaxies, and (4) the dark matter halos around and within the Virgo Supercluster.

For example, since the Milky Way is about 60 million light-years from the center of the Virgo Supercluster, it would not be surprising under relativistic dark matter theory, if only 3 to 4 protons per square kilometer per century would arrive at the Earth with energies at or above 10^{19} eV.

Note that the diameter of the Milky Way is about 100,000 light-years, its dark matter halo has an estimated outer diameter of about 1.5 million light-years, its Local Group of galaxies has an outer diameter of perhaps about 10 million light-years, and the outer diameter of the Virgo Supercluster is about 120 million light-years. Also note that the diameters of each of these celestial bodies is roughly about one order of magnitude greater than the one it encloses, which provides clues as to the kinetic energy levels of the orbiting protons in each case and how these energy levels relate to the energy levels of the "knees" and "ankle."

The relativistic proton dark matter theory leads to an estimate that the protons orbiting a cluster of spiral galaxies should have an average energy of about 30 times greater than those orbiting one of the spiral galaxies. (See Chapter 50, which arrives at this estimate of 30 using the DM halos around the Milky Way and around the Local Group of galaxies.)

TABLE 2

RECAP

Decoding The Cosmos Via DM Relationism And By Solving The Cosmic DM Mysteries

PART II – Cosmic DM Mysteries #1 - #14

#1	Spiral Disk Galaxies Have Spherical Dark Matter Halos
#2	Accelerating Expansion Via Conserving DM Momentum
#3	Hydrogen Derived From DM Cosmic Ray Protons
#4	Magnetic Fields Derived From DM Cosmic Ray Protons
#5	Intersecting DM Filaments Create Galaxy Clusters
#6	Mature Galaxies Discovered In The Very Early Universe
#7	Dark Matter Spherical Cored Halos Have "Hollow" Cores
#8	Source Of Spiral Galaxies'/Halos' Angular Momentum
#9	No Central Dark Matter Cusp Found In Spiral Galaxies
#10	LSB Dwarf Galaxies Have Low Star Formation Rates
#11	LSB Galaxies Have Inclining Star Rotation Curves
#12	Galaxy Hydrogen Is Replenished From Halo Dark Matter
#13	Dark Matter, Hydrogen, Helium, And Muons Create Stars
#14	Earthbound Cosmic Ray Protons Depart From 4 Locations

CHAPTER 43

SOME TENTATIVE CONCLUSIONS AFTER THE STUDY OF THE FIRST 14 OF THE 25 COSMIC DM MYSTERIES

Dark Matter Relativistic Protons Appear To Be A Much Stronger DM Candidate Than The Cold Dark Matter (CDM) Uncharged WIMPs And Neutralinos, For A Number Of Reasons:

By April 2005, the author had completed his study of the possible relationships between the relativistic proton dark matter theory/cosmology and 14 relevant Cosmic Constituents, now referred to as "Cosmic DM Mysteries." The material presented in this book through this Chapter 43 incorporates the results of that study.

At that point, the author was very encouraged and decided to search for more mysteries and unexplained or inadequately explained phenomena in the Universe that also could be designated Cosmic DM Mysteries. By October 2005, an additional eight relevant astrophysical or cosmological phenomena had been identified, which seemed to add credibility to the relativistic proton dark matter theory/cosmology. That was when the author decided to proceed with writing this book.

By January 2006, three additional relevant unexplained phenomena were discovered, bringing the total number of designated Cosmic DM Mysteries to 25. Chapters 44 through 54 present explanations for each of the 11 additional Cosmic DM Mysteries, utilizing only the relativistic proton dark matter theory/cosmology and the laws and principles of physics.

Tentative Conclusions Regarding The First 14 Of 25 Cosmic DM Mysteries:

- Orbiting dark matter relativistic protons appear to have an influence on or a relationship with at least 10 of the 14 Cosmic DM Mysteries compared to, at most, three to five for cold dark matter particles.

- Dark matter relativistic protons are detected every day as relativistic cosmic ray protons that bombard the Earth uniformly from all directions; whereas the theoretical CDM WIMPs and CDM neutralinos have never been detected during 20 years of searches following the announcement of the WIMP in 1984.

- Orbiting dark matter relativistic protons in collision with dust, hydrogen, helium, or CMB photons in space create muons and electrons that can transform the decelerating protons into hydrogen, the principal building block of ordinary matter.

- Dark matter relativistic protons appear to provide an explanation for the departing locations of Earthbound cosmic ray protons with energies at or above 10^{16} eV.

- Cluster-orbiting dark matter relativistic protons appear to provide a plausible explanation for an accelerating expansion of the Universe.

- The relativistic proton dark matter theory/cosmology, in conjunction with the Larmor Radius equation and the cosmic ray proton energy distribution at the Earth, provides explanations for the low SFRs for dwarf and LSB galaxies and the very high SFRs for large, massive, spiral galaxies for limited periods of time. See Chapters 23, 28, 38, 39, and 54.

- Multitudinous relativistic protons in Larmor orbits throughout the Universe are probably responsible for the galactic/extragalactic magnetic fields (whereas electrons in Larmor orbits would lose their kinetic energy too rapidly, through synchrotron radiation losses, to maintain the magnetic fields).

- Orbiting dark matter relativistic protons may have formed the spherical/ellipsoidal dark matter halos throughout the Universe through the principle of *astrophysical emergence* or *emergent evolution.* See Chapter 44. This *emergence* principle relies on strong self-interaction among the dark matter particles to achieve collective self-organization. The relativistic protons are strongly self interacting, meeting this *emergence* requirement, whereas the CDM *weakly interacting* massive particles do not appear to meet this requirement.

PART III

CHAPTER 44

Cosmic DM Mystery #15
Astrophysical Emergence Of Dark Matter Halos, After Eons

Astrophysical Emergence Of Dark Matter Halos And Long, Large, Dark Matter Filaments Could Place Constraints On The Identity Of Dark Matter Particles

The author would like to propose *astrophysical emergence* of dark matter halos as the 15th Cosmic DM Mystery of the Universe, which leads to the following rhetorical question: What type of dark matter particles could facilitate, expedite, or explain the formation of dark matter halos around galaxies and galaxy clusters through *astrophysical emergence?*

Is it possible that a simple microscopic entity or particle in enormous quantities could create, over an evolution period of hundreds of millions to a few billion years, dark matter halos around galaxies and galaxy clusters as well as long, large, dark matter filaments through the principle of *astrophysical emergence* or *emergent evolution?* Astrophysical emergence or emergent evolution or emergent behavior is involved when an enormous number of simple microscopic entities or particles interact with each other strongly so as to form a large entity of complex behavior through collective self-organization. The complex behavior of the resulting large entity would be unpredictable and unprecedented and would represent cosmic bodies or systems or Cosmic DM Mysteries with a high level of sophistication and evolution. These various characteristics are typical of *emergence.*

It is widely believed that *emergence* or emergent evolution of small, simple entities into a large entity of complex behavior requires that the simple pre-existing entities or particles be highly interactive with each other. Galaxy-orbiting relativistic protons meet this requirement.

Furthermore, the complex behavior of any such large entity would not be a known property of the simple microscopic entity or particle, nor could the complex behavior be deduced or predicted from the properties of the simple entity or particle. By relating *astrophysical emergence* to the formation of dark matter halos around galaxies and galaxy clusters and also to the formation of long, large, dark matter filaments that form galaxy clusters where they intersect, constraints can be placed on the nature of the simple microscopic entity (a dark matter particle) to help identify it.

When one deals with the concept of *emergence*, normally the focus is on the collective complex behavior of the resultant large, sophisticated entity because the characteristics of the simple entity or entities, present in enormous quantities, are known already. In those cases, the interest is in the unpredictable and unprecedented complex behavior of the large, sophisticated system that evolves.

The author has a different interest than that of traditional *emergence*. Most readers already are familiar with many of the various forms of complex behavior of the cosmic bodies and Cosmic DM Mysteries of the Universe such as galaxies, dark matter halos, galaxy clusters, the long, large filaments of dark matter, etc. In this case, the author is trying to identify the simple entity (or entities) that could have led to the collective self-organization and the complex behavior exhibited by these various cosmic bodies, systems, or Cosmic DM Mysteries.

If there is a principal simple entity or particle that collectively evolves into emergent behavior, it must have some special interactive characteristics. In the case of a hurricane, there are many simple entities interacting, such as pressure, temperature, moisture, wind, etc.

However, in *astrophysical emergence* involving dark matter halos, there are only a limited number of possible simple entities or particles that could qualify because, for example, they must also represent 80% to 85% of the mass of the Universe.

For example, even if the CDM WIMPs do exist, it is difficult to imagine that large numbers of *weakly interacting* massive particles (only through gravitation) could interact strongly enough with each other to exhibit *emergent* behavior -- namely, collective self-organization into a large cosmic body (such as the long, large filaments of dark matter) with complex behavior that is unpredictable and unprecedented.

On the other hand, the enormous numbers of relativistic protons interacting collectively, over an evolution period of hundreds of millions to billion years, possibly could form the

DM halos and the long, large, DM filaments that form galaxy clusters where they intersect. These DM filaments could represent one steady-state *emergence* solution. Following the crashing intersection of two DM filaments, new boundary conditions could be established that possibly could lead to the formation of spherical DM halos as a second steady-state *emergence* solution.

The relativistic protons individually follow a number of well-known principles of physics, which probably could lead to collective self-organization and complex behavior after an interaction period of hundreds of millions to a few billion years because of their strong and various coulomb, electromagnetic, and nuclear interactions listed below:

1. Relativistic protons create magnetic fields.

2. Relativistic protons are deflected (accelerated) by magnetic fields.

3. Relativistic protons follow almost-circular spiral paths that are determined by the Larmor Radius equation in direct proportion to the kinetic energies of the relativistic protons and inversely proportional to the strength of the local orthogonal magnetic field.

4. Relativistic protons emit synchrotron radiation and, as a result, lose kinetic energy and slow down, which leads to the emission of even more synchrotron radiation.

5. Relativistic protons moving from one level of orthogonal magnetic field to a higher one will experience an increase in synchrotron radiation.

6. When the velocity of a relativistic proton is reduced through synchrotron radiation losses or collisions with dust or atoms, relativistic mass is also reduced, thereby lowering its gravitational strength.

7. Collisions of relativistic protons with dust, molecules, or atoms will create pions that quickly decay into muons that, in turn, decay into electrons and thereby form a high-velocity plasma comprising protons, helium nuclei, electrons, and muons.

These seven characteristics of relativistic protons should play an important role in *astrophysical emergence*. If one is trying to explain the creation of DM halos and the long, large, DM filaments through the concept of *astrophysical emergence* utilizing either relativistic protons or weakly interacting theoretical WIMPs as possible simple microscopic entities, the relativistic protons with their strong and multiple interactive nature would be a far more promising candidate.

CHAPTER 45

Cosmic DM Mystery #16
UHECRs Arrive At Earth From Galaxy Superclusters

Ultra-High Energy Cosmic Ray Protons Arriving At Earth Probably Departed From A Galaxy Supercluster Or A Massive Galaxy Cluster

See SigChar A, C, D, G, M, and Chapters 46 and 47.

The author would like to propose ultra-high energy cosmic ray protons departing from massive galaxy clusters toward the Earth as the 16th Cosmic DM Mystery. The follow-up rhetorical question is: What type of dark matter particle could lead to ultra-high energy cosmic ray protons departing for Earth from massive galaxy clusters?

On April 22, 2005, astrophysics paper astro-ph/0504512[37] was posted by Drexler on the Physics arXiv. This astrophysics paper predicted that "The DM [dark matter] halos in and around the Virgo Supercluster probably would contain protons with energies above 6×10^{18} eV and probably would be the departing location of the highest energy cosmic rays arriving at the Earth." On July 29, 2005, astrophysics paper astro-ph/0507679 entitled, "Massive galaxy clusters and the origin of Ultra High Energy Cosmic Rays"[38] was posted on the Physics arXiv by astronomers Elena Pierpaoli (California Institute of Technology) and Glennys Farrar (New York University). Their paper reported:

> We find an excess of highest energy cosmic rays (above 50 EeV or 5×10^{19} eV) which correlate with massive galaxy cluster positions within an angle of about one degree. The observed correlation has a chance probability of order 0.1%.

Drexler's paper is entitled "Identifying Dark Matter Through the Constraints Imposed by Fourteen Astronomically Based 'Cosmic Constituents.'"[37] Although this 19-page paper discusses 14 Cosmic Constituents, (now referred to as "Cosmic DM Mysteries") of the Universe that may be created or influenced by or have a special relationship with possible dark matter candidates, only one of them involves the departure location of Earthbound ultra-high energy cosmic rays (UHECRs). This particular Cosmic DM Mystery is best described by the following rhetorical question: What type of dark matter particles could create the first "knee" at 3×10^{15} eV, the second "knee" between 10^{17}

eV and 10^{18} eV, and the "ankle" at 3×10^{18} eV of the cosmic ray energy distribution near the Earth? See Appendix B, Slide #17.

After determining that galaxy-orbiting *relativistic proton dark matter* best satisfies the constraints imposed by the 14 astronomically based Cosmic Constituents/Cosmic DM Mysteries, Drexler was then able to use this information to (1) explain the probable reasons for the two "knees" and the "ankle" of the cosmic ray energy distribution at the Earth and (2) identify four probable departing locations of Earthbound cosmic ray protons representing four different proton energy levels and associated proton flux levels.

This led to Drexler's conclusion in his April 22, 2005 paper that the departing location for Earthbound cosmic ray protons with energies greater than 6×10^{18} eV probably would be the dark matter halo in and around the Virgo Supercluster. In considering Pierpaoli and Farrar's July 29, 2005 paper, "Massive galaxy clusters and the origin of Ultra High Energy Cosmic Rays,"*[38]* note that a "massive galaxy cluster" would be an accurate description of the Virgo Supercluster; that about 90% of cosmic ray nuclei are high-velocity protons;

and that *ultra-high energy (UHE)* cosmic ray protons, also known as UHECRs, are defined as being at or above the 10^{18} eV energy level.

From the theoretical prediction of this book's author in April 2005 and from the actual astronomical detection several months later, it would appear that ultra-high energy cosmic ray protons arriving at Earth probably had departed from one or more massive galaxy clusters.

The explanations presented in this chapter for Cosmic DM Mystery #16 and the sources of the UHECRs are augmented by explanations and information presented in Chapter 47, for which Chapter 46 is a prerequisite.

CHAPTER 46

Cosmic DM Mystery #17
Starburst Galaxies Form Via Merging Galaxy Clusters

The Merging Of Spiral Galaxy Clusters Create Starburst Galaxies That Exhibit Star Formation Rates (SFRs) As Much As 50 Times Higher Than The SFRs Of Spiral Galaxies

See SigChar G, J, K, S, T, W, and X.

The author designates the above-titled phenomenon as the 17th Cosmic DM Mystery of the Universe, which then leads to the rhetorical question: What type of dark matter particle could cause, expedite, facilitate, or explain the starburst galaxy phenomenon?

Starburst galaxies show evidence of a transient increase in star formation rate by a factor as much as 50. Most starburst galaxies are associated with merging spiral galaxy clusters. The starburst phenomenon may be galaxy-wide or limited to a region of the galaxy such as the galaxy nucleus.

An excellent source of basic astronomical information about starburst galaxies is a six-page article, "Starburst Galaxies"[39] by astronomer William C. Keel of the University of Alabama, posted on the University of Alabama's website: http://www.astr.ua.edu/keel/galaxies/starburst.html.

Starburst galaxies, with SFRs rising by as much as a factor of 50, are created when two clusters of spiral galaxies merge. Starbursts are associated with spiral galaxies that have been disturbed from their steady-state condition. The starbursts are usually confined to a few hundred parsecs from the nucleus of a spiral galaxy, although some starbursts occur throughout the galaxy disk. Large quantities of dust are also associated with starburst galaxies as well as the blue stellar emission from the young stars. Much of the star formation appears to be associated with very compact star clusters of about one hundred million stars in a region of a few parsecs in diameter near the galaxy nucleus.

A majority of spiral galaxies that are found in close pairs, known as *interacting* galaxies, demonstrate an increase in SFRs from about 30% to is as much as 100%.

The four preceding paragraphs describe the principal features and characteristics of starburst spiral galaxies and interacting spiral galaxies. The next step is to utilize this information in conjunction with the relativistic proton dark matter theory/cosmology to extract a plausible explanation for the starburst phenomenon.

In the dark matter halo around a normal spiral galaxy, the lowest energy relativistic protons are close to and penetrate the surface of the enclosed galaxy, while the more than an order of magnitude higher energy relativistic protons would be orbiting at the outer diameter of the dark matter halo at radii more than an order of magnitude greater than those in close proximity to the galaxy. Therefore, there is normally no interaction between a galaxy's hydrogen atoms and molecules and the highest energy relativistic protons in the outer diameter of its dark matter halo.

However, if two spiral galaxies along with their relativistic dark matter halos are interacting or merging, the highest energy relativistic protons in the dark matter halo of one galaxy could bombard the hydrogen atoms and hydrogen molecules of the other galaxy. Further, if two spiral galaxy *clusters* are merging, the ultra-high energy protons orbiting the galaxy *clusters*, if disturbed from their steady-state orbits, can smash into the hydrogen atoms and molecules of individual spiral galaxies. The enormous number of muons generated by these UHECR collisions would be capable of catalyzing the hydrogen fusion reactions associated with the well-known starburst galaxy phenomenon.

If these last two paragraphs are not fully understood, it is suggested that the SigChar references at the beginning of this chapter be reviewed. SigChar T describes a low star formation rate dwarf galaxy where there is a physical gap between the galaxy disk and the "hollow" core of the halo, leading to a low star formation rate. It also explains how a normal star formation rate is associated with the disk of a spiral galaxy overlapping its dark matter halo "hollow" core, thereby permitting more and higher energy relativistic protons to enter the galaxy and generate muons where the hydrogen atoms and molecules are located. Applying these same concepts to closely situated interacting pairs of spiral galaxies, it is not surprising that their star formation rates rise about 30% to 100% above those of normal spiral galaxies.

Why should starburst galaxies have a factor of 50 higher star formation rate compared to the 30% to 100% rise for close and interacting pairs of spiral galaxies? There may be two reasons. Starburst galaxies are usually formed from the merging of two spiral galaxy *clusters*. The ultra-high energies of DM relativistic protons in the DM halo of a galaxy *cluster* are probably about 30 times higher than those in a DM halo of a single spiral galaxy. (See Chapter 50.) These ultra-high energy DM protons orbiting galaxy *clusters* have the potential to produce very high SFRs. Also, there should be a large reservoir of hydrogen atoms and molecules to ignite into stars near the spiral galaxy nucleus because gravitational tidal forces will move such slow-moving atoms and molecules toward the gravitationally attractive black hole over millions to billions of years.

The hydrogen atoms and molecules near the nucleus of a spiral galaxy would not normally participate in new star formation. Photographs of spiral galaxies typically exhibit the blue color of a new star formation at the outer periphery of those galaxies in the spiral arms. Therefore, it would not be surprising that in a starburst galaxy involving merging galaxy *clusters*, the ultra-high energy protons from the galaxy cluster's dark matter halo could be perturbed from their normal circular orbits because of magnetic field distortions and, thus, could smash into a spiral galaxy straight through to its hydrogen-rich nucleus to ignite new stars. Star ignition could occur through the creation of muons near the galaxy nucleus, followed by particle collisions involving hybrid muonic molecular ions comprised of protons and helium nuclei. Also facilitating this starburst process is the ionization of some of the atomic hydrogen gas in the galaxy nucleus by the ultra-high energy dark matter protons, which speeds up the formation of hydrogen *molecules* and thereby raises the SFR, as explained in Chapters 26 and 41.

This chapter provides a plausible explanation for the starburst galaxy phenomenon, designated Cosmic DM Mystery #17, that cannot be explained by the mainstream theory of star formation where clouds of hydrogen molecules collapse anywhere in a galaxy under their own weight and are heated through compression to hydrogen fusion temperatures.

CHAPTER 47

Cosmic DM Mystery #18
UHECR Protons Via Starburst Galaxies/Merging Galaxies

Spiral Galaxy Clusters, Merging To Form Starburst Galaxies, Were Recently Identified As A Source Of Ultra-High Energy Cosmic Ray Protons

See SigChar A, C, D, G, M, Appendix A, and also Chapters 45 and 46 and reference *39,* which should be considered an integral part of this chapter.

The author designates the above-titled phenomenon as the 18th Cosmic DM Mystery of the Universe, which raises the question: What type of dark matter particle could cause, expedite, facilitate, or explain the creation of ultra-high energy cosmic ray protons by merging galaxy clusters forming starburst galaxies?

In 2005, two research groups independently reported ultra-high energy cosmic rays (UHECRs) emanating from starburst-like galaxies, according to one group, or from merging galaxy clusters, according to the other group. Since starburst galaxies usually involve merging galaxy clusters, it would appear that both research groups could be on similar tracks to the same discovery. On May 13, 2005, Diego F. Torres and Luis A. Anchordoqui posted a paper on the Physics arXiv, astro-ph/0505283 entitled, "On The Observational Status Of Ultrahigh Energy Cosmic Rays And Their Possible Origin In Starburst-Like Galaxies."[40]

Then, on July 29, 2005, Elena Pierpaoli and Glennys Farrar posted a paper on the Physics arXiv, astro-ph/0507679 entitled, "Massive galaxy clusters and the origin of Ultra High Energy Cosmic Rays,"[38] in which the "massive galaxy clusters" are described as "a merging pair of clusters." In their paper, Pierpaoli and Farrar suggest a possible explanation for the observed phenomenon as follows:

> A merging pair of clusters would be expected to have very large scale, strong magnetic shocks which could be responsible for accelerating UHECR even if there is no AGN [active galactic nucleus] or GRB [gamma ray burst] associated with the galaxy clusters.

Note that Pierpaoli and Farrar believe that lower-energy cosmic ray protons are accelerated into UHECRs through magnetic shocks created in the merging galaxy *clusters*. The paper by Torres and Anchordoqui similarly concludes that lower energy cosmic ray protons are probably accelerated to ultra-high energy UHECRs within the merging starburst-like galaxies.

Perhaps both research groups are correct in concluding that the UHECRs may have been accelerated. However, there is another possibility. During the pre-merger period, UHECRs, defined as having energies at or above 10^{18} eV, might have been orbiting galaxy *clusters* within their dark matter halos in a steady-state manner according to the Larmor Radius equation. Given the general size of galaxy clusters and the generally accepted magnitude range of the extragalactic magnetic field, one would conclude that most of the pre-merger orbiting protons in the dark matter halos around the galaxy clusters would be UHECRs.

The galaxy cluster merging process would upset the steady-state Larmor orbiting symmetry of the UHECRs. The combining of the magnetic fields of the two merging spiral galaxy clusters could create transient magnetic field distortions, which would cause a number of UHECRs to be deflected off into space, with some being Earthbound. This theory might be called the deflection-from-orbit theory of UHECR emission. It is presented as a plausible alternative theory to the shock acceleration UHECR theory, which remains unproven according to the two research groups.

The Pierpaoli and Farrar paper states that the authors have data of UHECRs with energies at $5x10^{19}$ eV departing from a merging pair of galaxy clusters observed in the SSDS DR3. In the Torres and Anchordoqui paper, the authors indicate the energies of the UHECRs emanating from a starburst galaxy are at or above 10^{18} eV.

Both of these sets of astronomical data appear to be consistent with Drexler's proposed deflection-from-orbit theory of UHECR emission. Therefore, this chapter provides a plausible explanation for the merging-galaxy-cluster-origin-of-UHECRs phenomenon defined by Cosmic DM Mystery #18, as per the Pierpaoli and Farrar paper and possibly as per the Torres and Anchordoqui paper, provided that their starburst galaxy evolved from merging spiral galaxy clusters.

CHAPTER 48

Cosmic DM Mystery #19
Blue Stars In Spiral Arms Vs. Red Stars In Galaxy Nucleus

The Spiral Arms Of Spiral Galaxies Contain Many Hot Blue And Blue-White Stars Less Than One Million Years Old, And In The Galaxy Nucleus There Are Red Stars About Five Billion Years Old

See SigChar A, C, D, G, J, T, W, X, and Chapters 31 and 38.

The author designates the above-stated phenomenon as the 19th Cosmic DM Mystery of the Universe, which raises the rhetorical question: What type of dark matter particle could cause, expedite, facilitate, or explain the creation of the one-million year-old blue and blue-white stars in the spiral arms at the outer diameter of a spiral galaxy, which also has five-billion-year-old red stars located near the galaxy nucleus?

A study of face-on photographs of spiral galaxies clearly shows the many blue and blue-white young stars in the spiral arms at the outer periphery of spiral galaxies. A photograph of the Andromeda galaxy, M31, can be found on the cover of this book. Photographs directed toward the nuclei of spiral galaxies at red wavelengths show old red stars that are estimated to be five billion years old (for Galaxy M81).

These star colors and their locations are widely known. Although there may be generally accepted explanations for the location of the newborn blue and blue-white stars in the spiral arms at the outer periphery of spiral galaxies, the author has not been able to find any published explanation for this preferred star-birth location. This specific star-birth location is not explained by the mainstream theory of star formation where clouds of hydrogen collapse, anywhere in a galaxy, under their own weight and are heated, through compression, to hydrogen fusion temperatures. The mainstream star formation theory provides no clues why spiral galaxies should form their new stars in the spiral arms.

Why are the blue and blue-white stars being formed at the outer diameter of the spiral galaxy disks M81and M31 rather than elsewhere? Could the spherical dark matter halos surrounding spiral galaxies such as M81 and M31, the Andromeda galaxy, play a role in igniting new stars in them?

Let us briefly review the steps in achieving new star formation in an isolated spiral galaxy according to the relativistic proton dark matter star formation theory. This review also explains why the stars forming in an isolated spiral galaxy, surrounded by a relativistic proton dark matter halo, would be located near the outer diameter of the galaxy disk, where the spiral arms are located:

1. An isolated spiral galaxy is surrounded by a spherical dark matter halo comprised of galaxy-orbiting relativistic protons following almost-circular spiral paths determined by the proton kinetic energies, the local orthogonal magnetic field, and the Larmor Radius equation.

2. Since a spiral galaxy is normally producing stars, its disk must be overlapping the "hollow" core diameter of the dark matter halo. (Spherical dark matter halos surrounding spiral galaxies have "hollow" cores, as explained at the end of Chapter 38.) Relativistic protons near the "hollow" core would lose kinetic energy over time from synchrotron radiation losses and eventually would plunge into the enclosed spiral galaxy and bombard the atomic and molecular hydrogen gas closest to the galaxy surface.

3. The bombarding of the hydrogen gas primarily near the surface by the relativistic protons will produce a number of significant effects. By ionizing some (50% would be optimum) of the atomic hydrogen, the conversion of

atomic hydrogen to molecular hydrogen would be accelerated. The protons bombarding the resulting molecular hydrogen would create large numbers of muons, which would catalyze hydrogen fusion reactions. The bombarding protons (and accompanying high-speed helium nuclei) also would trigger nuclear fusion reactions and the ignition of new stars by colliding with the muonic ions created earlier by reactions of muons with the hydrogen and helium.

This scenario provides a plausible explanation for the birth of blue and blue-white stars in the spiral arms near the surfaces of spiral galaxies, as defined by Cosmic DM Mystery #19.

The dark matter relativistic protons also may be performing another role. On one hand, the muons they create can catalyze the hydrogen fusion reaction, thereby igniting new blue and blue-white stars. In addition, the protons add hydrogen to the enclosed galaxy, which can facilitate future star formation and to cause the galaxy to grow by accretion.

The idea of galaxy growth through accretion of baryons provided by the galaxy's dark matter halo is a relatively new idea and, therefore, requires some support from astronomical data. One example that provides such support is the

Andromeda galaxy, M31, whose disk seems to have enlarged by accretion by a factor of three over billions of years. See astro-ph/0504164 entitled, "On the accretion origin of a vast extended stellar disk around the Andromeda galaxy."[41]

There is a related point that should be considered. Five billion years ago, the red stars near Andromeda's nucleus might have been blue stars at the outer diameter of a then-much-smaller galaxy. Did the galaxy disk grow in diameter by accretion since then?

A study of 2,800 stars outside Andromeda's disk by the authors of astro-ph/0504164, led to the discovery that these stars were not in the halo of the disk, but actually in an extension of the galaxy disk. From a study of the velocities and directions of the stars in the extended disk, researchers have ruled out the possibility of an earlier merger with another galaxy to explain the tripling of the disk diameter. This would leave hydrogen accretion as the primary source of growth, by a factor of three, in the diameter of the Andromeda disk. The author believes that Andromeda's dark matter halo could have provided the necessary hydrogen or protons for galaxy disk growth as posited by the relativistic proton dark matter theory/cosmology; however, astro-ph/0504164 does not suggest that.

Additional astronomical support for new star formation taking place primarily at the outer periphery of spiral galaxies is found in the empirical Schmidt law that posits that the SFR of isolated spiral galaxies is highly correlated with their average *molecular hydrogen surface density*.

See Chapter 53, which links the Schmidt law to the relativistic proton dark matter theory/cosmology and thereby provides (1) a plausible explanation for the Schmidt law, (2) support for the relativistic proton dark matter theory and cosmology, and (3) support for Cosmic DM Mystery #19, the formation of new stars in the spiral arms near the surfaces of spiral galaxies.

The star formation phenomena defined by Cosmic DM Mystery #19 cannot be explained by the generally accepted mainstream theory of star formation where clouds of hydrogen molecules collapse anywhere in a galaxy under their own weight and are heated through compression to hydrogen fusion temperatures.

CHAPTER 49

Cosmic DM Mystery #20
Magnetic Field, DM Proton Energies Set Galaxy Halo Size

The Only Dark Matter Particle Candidate That "Predicts" The Size Of The Milky Way's DM Halo Is The Relativistic Cosmic Ray Proton Moving In The Extragalactic (Intergalactic) Magnetic Field Having A Strength Of About 1×10^{-9} Gauss

See SigChar A, B, C, D, and M.

The author designates the above-titled phenomenon as the 20th Cosmic DM Mystery of the Universe, which raises the question: What type of dark matter particles could orbit the Milky Way's 30.7 Kpc diameter (15.4 Kpc radius) disk within its dark matter halo over a range from a radius of about 11 Kpc at the inner diameter of the "hollow" core of its dark matter halo to a radius of about 225 Kpc at the outer diameter of its dark matter halo?

The diameter of the Milky Way galaxy disk is about 100,000 light-years or 30.7 Kpc. Astronomers have estimated that the dark matter halo of a spiral galaxy extends to about 10 to 20 times the size of the visible regions. Using a factor of 15, the outer radius of the dark matter halo would extend to perhaps 225 Kpc.

Astronomers also have estimated that the dark matter halo overlapping the Milky Way's disk represents a dark matter mass penetration about equal to the mass of the Milky Way.

The Larmor Radius for a 10^{16} eV proton in the Milky Way halo's extragalactic magnetic field of 1×10^{-9} gauss is 11 Kpc; for a 10^{17} eV proton, it is 110 Kpc; and for a 10^{18} eV proton, it is 1,100 Kpc.

Therefore, at the dark matter halo's outer radius of 225 Kpc from the Milky Way's nucleus, the relativistic proton energy would be estimated at about 2×10^{17} eV using the Larmor Radius equation, while the proton energy at the 11 Kpc inner core radius of the DM halo would be estimated at about 1×10^{16} eV.

This range of proton energy levels, based upon the Larmor Radius equation, is surprisingly consistent with the proton energy range of 1×10^{16} to 3×10^{17} based upon the cosmic ray proton energy levels for the dark matter halo surrounding the Milky Way. See Chapter 16 and the graph of cosmic ray [proton] energy distribution at the Earth (Appendix B, Slide #17).

Cosmic DM Mystery #20 is defined by the author, as follows: The only dark matter particle candidate that "predicts" the size of the Milky Way's DM halo is the galaxy-orbiting relativistic cosmic ray proton moving in the extragalactic (intergalactic) magnetic field, which has a field strength of about 1×10^{-9} gauss.

The author achieved the above results by utilizing astronomers' estimates for the size of the Milky Way and its dark matter halo, the strength of the extragalactic magnetic field, the penetration of the DM halo into the Milky Way, and the cosmic ray proton energy distribution at the Earth, in conjunction with the use of the Larmor Radius equation.

The author would like the reader to examine the author's definition of the term *relationism* (see below) to determine whether this chapter's presentation is an example of the effective use of *dark matter relationism* as an analytical method to facilitate the identification of dark matter.

> **Relationism**: An analytical procedure, method, concept, or theory, developed by Jerome Drexler, that attempts to identify dark matter by determining which cosmic phenomena (called Cosmic DM Mysteries) may be facilitated, expedited, influenced by, or have a special relationship with dark matter. This outward-looking cosmological approach attempts to determine the nature and characteristics of dark matter's influence on and relationship with many different cosmic phenomena as a means of identifying dark matter.

In summary, the relativistic proton dark matter theory and cosmology appear to be compatible with (1) astronomers' estimates for the size of the Milky Way and its dark matter halo, (2) the strength of the extragalactic magnetic field, (3) penetration of the DM halo into the Milky Way, and (4) the cosmic ray proton energy distribution at the Earth.

CHAPTER 50

Cosmic DM Mystery #21
Different Dark Matter For Small Galaxies And For Clusters

Two Different Types Of Dark Matter Halo Particles Reported For Smaller Galaxies And For Galaxy Clusters. On August 26, 2005, Researchers Reported That In Low-Surface Brightness Galaxies And Dwarf Galaxies, The Self-Interaction Cross Section Of Dark Matter Particles Appears To Be High; While For More Massive Systems Such As Galaxy Clusters, The Self-Interaction Cross Section Of Dark Matter Particles Appears To Be Low.
No Explanatory Theory Was Offered.

See SigChar C, D, G, M, S, and Chapter 44.

The author selects the above-stated dual-dark-matter phenomenon to be the 21st Cosmic DM Mystery of the Universe, which raises the question: What type of dark matter particle could have a high self-interaction cross section for DM halos of low surface brightness galaxies and dwarf galaxies, but a lower self-interaction cross section for DM halos of spiral galaxy clusters?

On August 26, 2005, Bo Qin, Ue-Li Pen, and Joseph Silk posted a four-page paper, "Observational Evidence for Extra Dimensions from Dark Matter,"[42] on the Physics ArXiv as astro-ph/0508572. They pointed out that the cold dark matter theory was questioned in the late 1990s because numerical simulations indicated a cuspy core, which was never observed astronomically. They then gave credit to David N. Spergel and Paul J. Steinhardt of Princeton University, the authors of astro-ph/9909386, "Observational evidence for self-interacting cold dark matter,"[43] for proposing in 1999 a self-interacting cold dark matter mode that could possibly solve the cuspy core problem.

However, astro-ph/0508572 points out that in the year 2000, numerical simulations:

> ... found that the [self-interaction] cross sections needed to produce good agreement with galaxies turned out to produce galaxy clusters too large or too round to be consistent with observation.

Bo Qin, Ue-Li Pen, and Joseph Silk then came to a very significant conclusion:

> These lines of evidence might suggest that, instead of being a fixed value, the dark matter self-interaction cross section

as proposed by Spergel and Steinhardt, is likely to vary in different dark matter systems -- smaller systems (like dwarf galaxies) tend to have larger cross sections whereas more massive systems (like galaxy clusters) tend to have smaller cross sections.

These researchers then provided an equation proposed by others that estimates the self-interaction cross section of dark matter particles. They then added the following:

> The nature of this self-interaction between dark matter particles is unknown. Its strength generally must be put in by hand. We note that the scattering cross section in Eq.(1) decreases with increasing velocity, which is characteristic of long-range forces (like gravity or Coulomb forces).

The above-quoted three paragraphs essentially describe the characteristics of the relativistic protons in the relativistic proton dark matter theory. Chapter 49 points out that the proton energies in the Milky Way's dark matter halo probably range in energy between 1×10^{16} eV and 2×10^{17} eV, while Chapter 16 points out the proton energies in the dark matter halo surrounding the Local Group cluster of galaxies probably range in energy levels between 3×10^{17} eV and 6×10^{18} eV, or about 30 times higher.

Note that this book is dealing with the coulomb forces of the protons and, therefore, the self-interaction or interaction cross section declines with proton velocity or as the square root of energy. Using this information and the two previous paragraphs about spiral galaxies and their clusters, one can arrive at a roughly estimated factor of about five higher self-interaction cross section for the relativistic protons in the DM halos of dwarf and LSB galaxies, as compared with the relativistic protons in the DM halos of spiral galaxy clusters.

Thus, the relativistic proton dark matter theory appears to be compatible with the dual-dark-matter phenomenon defined by Cosmic DM Mystery #21, for which a plausible explanation has been provided in this chapter. This is an indication that *dark matter relationism* seems to be applicable to Cosmic DM Mystery #21.

Chapter 44 indicates the need for the dark matter particles to be highly self-interactive in order that *emergent evolution* principles could lead to the formation of dark matter filaments and dark matter halos around galaxies/clusters during an evolution period of millions to a few billion years.

CHAPTER 51

Cosmic DM Mystery #22
800 Galaxies Detected, Less Than 1.2 Billion Years Old

Report Of Over 1,000 Clumps Of Dark Matter, With Most Harboring Several Newborn Galaxies, 12 Billion Light-Years Away. On December 21, 2005, Japanese Astronomers Reported That Clumps Of Dark Matter Are The Nursing Grounds For 5, 000 Newborn Galaxies About 12 Billion Light-Years Away. A Single Nest Of Dark Matter Can Nurture Several Young Galaxies. About 800 Of The Galaxies Are 12.5 Billion Light-Years Away And, Therefore, Were Born Less Than 1.2 Billion Years After The Big Bang.

See SigChar O, P, Q, and W.

The above-indicated astronomical data and the December 21, 2005 research report entitled, "Young Galaxies Grow Up Together in a Nest of Dark Matter,"[44] represent the work of researchers at the Space Telescope Institute, the National Astronomical Observatory of Japan, and the University of Tokyo.

The author of this book chooses the above-reported multiple-galaxy nurseries to be the 22nd Cosmic DM Mystery of the Universe, which raises the question: What type of dark matter particles could form enormous clumps of dark matter in the early Universe only 1.2 billion years after the Big Bang, could nurture several young galaxies simultaneously, and could also play a role in their initial galaxy formation? The galaxy-orbiting relativistic proton dark matter seems to have several necessary characteristics to satisfy the requirements of Cosmic DM Mystery #22.

First of all, owing to their velocities, relativistic proton dark matter particles could have formed a very large number of widely distributed, enormous clumps of dark matter within 1.2 billion years after the Big Bang. That is, their relativistic velocities enable them to quickly distribute the dark matter over a very wide area. More specifically, with the filamentary dark matter structure moving at, say, 90% of the speed of light along, say, a circular path, the protons could have encompassed a circular area of dark matter clumps with a diameter of roughly about 300 million light-years, by the end of the 1.2-billion-year evolutionary period.

Secondly, relativistic proton dark matter can form several new galaxies within each of the widely distributed, very large clumps of dark matter by means of the top-down galaxy formation process, which is an integral part of relativistic proton dark matter theory/cosmology. This takes place in several steps. Protons that are slowed through synchrotron radiation and collisions, combine with electrons to form hydrogen, the basic ingredient of galaxies. (These electrons were created when the relativistic protons collided with hydrogen and helium atoms and molecules, with dust and with photons to create pions initially, which rapidly decayed into muons that then decayed into electrons.)

This formation of hydrogen from decelerated protons and electrons is a key step in the formation of the new galaxies within the dark matter clumps. By this means, the dark matter clumps can accrete hydrogen onto the enclosed proto-galaxies, causing them to grow. The fact that several newborn galaxies are located in a single clump of dark matter provides an image of a top-down galaxy formation process, where the dark matter clump forms first and the galaxies form afterward within it and grow by the accretion of protons and hydrogen, derived from dark matter protons converted to hydrogen.

Finally, relativistic proton dark matter can form stars in the newborn galaxies by bombarding the hydrogen and helium atoms and molecules with high-velocity protons to generate muons, which then form muonic atoms, molecules, and ions. In turn, these muons, muonic ions, molecules, and atoms can catalyze hydrogen fusion reactions and can trigger the birth of new stars by the subsequent proton/helium nuclei bombardment. See Chapter 26, SigChar W, which discusses the first-generation star formation process.

Based upon the above, relativistic dark matter protons appear to meet the requirements defined by Cosmic DM Mystery #22. They have the relativistic velocities to cause the rapid distribution of dark matter over regions hundreds of millions of light-years in size; they can create atomic hydrogen for use in forming galaxies; they can facilitate hydrogen molecule formation through ionization of some of the atoms; and they also can form and ignite stars by creating muons, then muonic ions, and then subjecting the muonic ions to high-velocity proton/helium nuclei bombardment. Meeting these requirements is an indication that *dark matter relationism* is applicable to Cosmic DM Mystery #22.

CHAPTER 52

Cosmic DM Mystery #23
Fine Balance Between Dark And Baryonic Matter In Spirals

On September 12, 2005 An American Astronomer, Stacy McGaugh, Reported "A Fine Balance Between Dark And Baryonic Mass Is Observed In Spiral Galaxies. As The Contribution Of The Baryons To The Total Rotation Velocity Increases, The Contribution Of The Dark Matter Decreases By A Compensating Amount."

See SigChar A, G, J, L, O, W, X, and Chapter 31.

The above-indicated astronomical data was posted on the Physics arXiv on September 12, 2005 as astro-ph/0509305, entitled, "The Balance of Dark and Luminous Mass in Rotating Galaxies,"[45] by Professor Stacy McGaugh of the University of Maryland.

The above-quoted "A fine balance between dark and baryonic mass is observed in spiral galaxies" is chosen as the 23rd Cosmic DM Mystery of the Universe. This raises the rhetorical question: What type of dark matter particles could establish a fine balance between dark and baryonic mass within a group of spiral galaxies, such that when the contribution of the baryons to the total rotation velocity is seen to be larger in a specific spiral galaxy, the contribution of the dark matter is seen to be smaller by a compensating amount in the specific galaxy?

The author believes that Professor McGaugh may have made an important dark matter discovery; and, therefore, portions of his astro-ph/0509305 paper are presented here in their original form to ensure that his discovery is accurately presented:

Astrophysics, abstract
astro-ph/0509305

From: Stacy McGaugh [view email]
Date: Mon, 12 Sep 2005 20:15:55 GMT (46kb)

The Balance of Dark and Luminous Mass in Rotating Galaxies

Authors: Stacy McGaugh (University of Maryland)

Comments: 4 pages RevTeX. Phys. Rev. Letters, in press
Journal-ref: Phys.Rev.Lett. 95 (2005) 171302
A fine balance between dark and baryonic mass is observed in spiral galaxies. As the contribution of the baryons to the total rotation velocity increases, the contribution of the dark matter decreases by a

compensating amount. This poses a fine-tuning problem for ΛCDM galaxy formation models, and may point to new physics for dark matter particles or even a modification of gravity.

At the end of his paper, Professor McGaugh sought explanations of the *dark matter/baryon balance effect* and reviewed possible explanations, as follows:

> There is a third possibility. We still know very little about the nature of the dark matter (presuming it exists). It may possess some property that imparts the observed balance with baryons in galaxies. This idea implies a specific interaction between the two that is in some way repulsive: the greater the surface density of baryons, the less that of dark matter. Such a repulsion would help explain the apparent lack of dark matter in high density regions like globular clusters [45] and elliptical galaxies [46]. It would also have important implications for direct detection experiments.
>
> No interaction with baryons of the sort envisaged is in the nature of most hypothesized dark matter candidates. Neither cold dark matter nor frequently discussed alternatives like warm [47, 48] or self-interacting [49] dark matter do anything of the sort. Whether it is even possible to endow dark matter with the appropriate properties [50] is difficult to say as the possibility has yet to be thoroughly explored.

After reading Professor McGaugh's paper, Drexler sent him the following email. It provides an explanation of how the galaxy-orbiting relativistic proton dark matter could achieve the observed phenomenon (Cosmic DM Mystery #23) as described in the Abstract of Professor McGaugh's paper, namely: "A fine balance between dark and baryonic mass is observed in spiral galaxies. As the contribution of the baryons to the total rotation velocity increases, the contribution of the dark matter decreases by a compensating amount."

To: Professor Stacy S. McGaugh, Astronomy Department, University of Maryland

From: Jerome Drexler, Research Professor, Department of Physics, New Jersey Institute of Technology, Los Altos Hills, California

I read your dark matter paper, Astro-Ph/0509305, with great interest. It goes a long way toward supporting a dark matter candidate I have been proposing for several years. My dark matter paper, astro-Ph/0504512, entitled, "Identifying Dark Matter Through the Constraints Imposed By Fourteen Astronomically Based 'Cosmic Constituents'" provides the rationale for relativistic-proton dark matter that may satisfy the DM constraints suggested by your paper.

My paper points out that the dark matter halo enclosing spiral galaxies is comprised of galaxy-orbiting relativistic protons moving in spiral paths determined by the extragalactic/galactic magnetic fields and the Larmor Radius equation.
In my top-down theory, the galaxy hydrogen is derived from slowed dark matter halo

protons in conjunction with electrons created from decaying muons, which were produced by collisions of the orbiting relativistic dark matter halo protons with dust, CMB photons, and hydrogen and helium.

Therefore, there appears to be a natural inverse relationship between the mass of the dark matter halo and the mass of the galaxy baryons in spiral galaxies.

That is, as the DM halo protons are slowed and collect electrons, they leave the DM halo thereby reducing its mass and move into the enclosed galaxy and become new hydrogen in the galaxy, thereby increasing its mass.

I hope you find my paper and DM candidate of interest and value.

Note that Professor McGaugh points out in his abstract (see above), "This poses a fine-tuning problem for ΛCDM [cold dark matter] galaxy formation models."

The author of this book hopes that the reader will find Drexler's above e-mail explanation regarding Professor McGaugh's dark-matter-fine-balance phenomenon, represented by Cosmic DM Mystery #23, to be plausible. That would indicate *dark matter relationism* is applicable to Cosmic DM Mystery #23.

CHAPTER 53

Cosmic DM Mystery #24
Schmidt Law: SFR Vs. Surface Hydrogen Molecular Density

One Of The Mysteries Of Observed Isolated Spiral Galaxies Has Been The Empirical Schmidt Law Correlation Between Star Formation Rate And The Average Molecular Hydrogen Surface Density On Kiloparsec Scales, Even In Regions Dominated By Atomic Hydrogen. The Schmidt Law Has A Power-Law Index Of Correlation Or Slope In The Range Of 1.3 To 1.5. At High Gas Densities, The Schmidt Law Is Very Consistent From Galaxy To Galaxy But Does Break Down Below A Threshold Surface Density Level.

See Chapters 23, 26, 27, 31, 41, 46, and 48 and References *46* (astro-ph/0303240, "On the Origin of the Global Schmidt Law of Star Formation") and *47* (astro-ph/0508054, "Star Formation in Isolated Disk Galaxies. II. Schmidt Laws and Star Formation Efficiency of Gravitational Collapse").

The author chooses the Schmidt law for isolated spiral galaxies to be the 24th Cosmic DM Mystery of the Universe. That raises the question: What type of dark matter particles could cause, facilitate, expedite, or explain the Schmidt law for isolated spiral galaxies?

The Schmidt law appears to have some distinctive characteristics:

1. It is a power law with a slope (of 1.3 to 1.5) almost half way between a linear and a square law correlation.

2. Although the law is a strong function of the average molecular hydrogen surface density, it does not seem to be affected by the level of surface atomic hydrogen.

3. At high levels of molecular hydrogen surface density, the law is consistent from galaxy to galaxy.

4. The empirical Schmidt law uncovered a molecular hydrogen surface-density-feature correlation as distinct from a volume-density-feature correlation.

5. The law breaks down and therefore is not valid below a low molecular hydrogen surface density threshold level.

6. The law does not apply to starburst galaxies, which normally involve merging galaxy clusters and therefore are not isolated spiral galaxies.

Using the relativistic proton dark matter theory/cosmology, we will now seek a plausible explanation for the Schmidt law. Success in doing so will confirm that *dark matter relationism* is applicable to the Schmidt law, which is the subject of Cosmic DM Mystery #24.

A review of Chapters 26, 27, 31, and 41 should provide the background information necessary to understand the star formation and hydrogen fusion process associated with the relativistic proton dark matter theory/cosmology.

The relativistic proton dark matter theory/cosmology's star formation process for an isolated spiral galaxy involves four steps, as follows:

1. Relativistic protons with energies at or below 10^{16} eV depart from the dark matter halo in closest proximity to the spiral galaxy and move into the galaxy as a result of proton energy losses from synchrotron radiation losses and collisions with CMB photons, helium, hydrogen, and dust.

2. When these relativistic protons collide with the hydrogen molecules near the spiral galaxy's surface, they generate pions that quickly decay into muons that decay less quickly into electrons. Hundreds to one thousand muons can be generated by each 10^{15} eV proton.

3. These large quantities of muons then collide with the hydrogen molecules near the surface of the spiral galaxy as well as with the estimated one-in-twelve helium molecules present on the galaxy surface. As a result,

muonic hydrogen ions and muonic hydrogen-helium ions are created near the galaxy surface.

4. The relativistic protons arriving at the spiral galaxy surface after the muonic ions are formed, bombard the muonic hydrogen ions and the muonic hydrogen-helium ions, thereby initiating hydrogen fusion reactions. (Note that these bombarding protons are about 1,000 times more energetic than man-made relativistic protons.)

It should be noted that the dark matter halo is also providing replenishment hydrogen fuel to the galaxy as the slowed protons entering the galaxy will capture some of the enormous quantity of electrons being generated by the decaying muons. Also, it is known that muons usually catalyze multiple hydrogen fusion reactions during their short lifetime since they are not destroyed in the nuclear fusion process and, therefore, would be in plentiful supply even though they decay rapidly into electrons.

Drexler first published this hydrogen fusion star ignition theory on April 22, 2005 on the Physics arXiv in astro-ph/0504512.[37] This star ignition theory is based upon and relies on the relativistic proton dark matter theory and its related cosmology. We will now apply Drexler's star formation/ignition theory to the characteristics of the

Schmidt law to determine whether Drexler's theory provides a plausible theoretical basis and explanation for the empirical Schmidt law.

Why should the Schmidt law be a power law? Note that the star forming/ignition process involves four steps, two of which involve relativistic proton bombardment. That is, the higher the average hydrogen molecular surface density, the more muons will be created per bombarding proton. Then, for a given number of muons created, the higher the hydrogen molecular surface density, the more hydrogen and helium molecules at the surface will be converted into muonic ions. Finally, the arriving relativistic protons bombarding the muonic ions will create a level of star formation in proportion to the surface density of muonic ions. Thus, the posited star formation process should have, and appears to have, a somewhat stronger than linear correlation with the average molecular hydrogen density on the surfaces of isolated spiral galaxies, which at the least would imply a low-level power law.

The author believes this chapter provides a plausible theoretical basis and explanation for the empirical-Schmidt-law phenomenon defined by Cosmic DM Mystery #24. This is an indication that *dark matter relationism* is applicable to Cosmic DM Mystery #24.

Why doesn't the presence of atomic hydrogen affect the correlation of the Schmidt law? First of all, the level of muon creation for atomic hydrogen would be much lower than that for hydrogen molecules, which have a larger collision cross section to the relativistic proton/helium nuclei bombardment. Also, in the atomic-hydrogen case, the muons created will not produce the large numbers of muonic ions required for hydrogen fusion since there would be relatively small numbers of hydrogen molecules, helium molecules, and hydrogen-helium molecules present near the galaxy surface to convert to the muonic ions.

These star formation phenomena defined by Cosmic DM Mystery #24 cannot be explained by the generally accepted mainstream theory of star formation where clouds of hydrogen molecules collapse anywhere in a galaxy under their own weight and are heated through compression to hydrogen fusion temperatures.

CHAPTER 54

Cosmic DM Mystery #25
Mass-And-Time-Dependent SFR Graphs For Field Galaxies

The Two-Part Mystery Of Recently Observed Star-Forming Galaxies Is That Large, Massive Galaxies Form Stars Early And Rapidly, But Eventually Their Star Formation Rates Fall Rapidly, Whereas Small Galaxies Form Stars Slowly Over Longer Time Scales And Their Star Formation Rates Decline Slowly Over Longer Time Scales

See Chapters 23, 26, 27, 28, 31, 41, 46, and 48.

For the 25th Cosmic DM Mystery of the Universe, the author chooses the discovery that time-dependent star formation rates differ significantly for massive star-forming galaxies compared to small star-forming galaxies. This raises the question: What type of dark matter particles could cause, facilitate, expedite, or explain the mass-and-time-dependent SFR phenomena exhibited by massive star-forming galaxies and by small star-forming galaxies?

Mass-dependent SFR phenomena for star-forming galaxies were reported on December 18, 2005 in a research paper, astro-ph/0512465, "The Mass Assembly History of Field Galaxies: Detection of an Evolving Mass Limit for Star Forming Galaxies,"[48] by Kevin Bundy, Richard S. Ellis, et al. A similar subject was presented on January 9, 2006 at the AAS 207th meeting in Washington, D.C., in a talk entitled, "Mass-dependent star formation histories of field galaxies in the EGS" by Kai G. Noeske, which was reported in a January 9, 2006 University of California at Santa Cruz news release entitled, "Large survey of galaxies yields new findings on star formation."[49]

The above-referenced paper and news release provide the following information about the SFR as a function of star-forming galaxy mass and as a function of time:

- Massive, undisturbed, star-forming galaxies form stars early and rapidly.

- Small, undisturbed, low mass star-forming galaxies form their stars more slowly and over longer time scales than the massive galaxies.

- Both massive and small star-forming galaxies exhibit a decline in star formation rate over time, but the fall of the

SFR for the massive galaxies is much more rapid and begins much earlier.

- Astronomical studies of large galaxy clusters representing billions of years of star formation history demonstrate a "downsizing signal." This means a "downsizing" pattern in space and time, in which the sites of the most active star formation shift over time in a large galaxy cluster from its massive galaxies in the early epochs to its low mass galaxies during later epochs.

- The fading of star formation rates over time appears to be consistent with galaxies exhausting their hydrogen gas, rather than for any other causes.

These mass-and-time-dependent SFR phenomena seem to be explainable by these paragraphs from Chapters 28 and 23:

From Chapter 28: Under the top-down theory of galaxy formation, galaxies form and grow through the accretion of hydrogen and protons from a relativistic proton DM halo into its "hollow" core and onto its enclosed proto-galaxy. The galaxy disk enclosed within a DM halo may be larger or smaller than the relativistic proton DM halo's "hollow" core diameter. If the galaxy disk is smaller, the galaxy should exhibit a low SFR; and if the disk is much larger, such a massive galaxy should exhibit a very high SFR. The decline of SFRs over time would be very small for small galaxies because their DM halos could provide adequate replenishment hydrogen to compensate for the hydrogen consumption of its stars. However, for large massive galaxies with their very high SFRs, the replenishment hydrogen from the DM halos could be inadequate, subjecting the SFRs to declines over time that are large and rapid.

From Chapter 23: An unusually large, massive galaxy could significantly overlap the DM halo's "hollow" core, leading to a very high SFR. Such a massive galaxy's disk would be reaching deep into the DM halo, thereby interacting with higher energy relativistic protons and thus creating a very active high SFR region. However, if this SFR is too high, the replenishment hydrogen being supplied by the DM halo could be insufficient to maintain the SFR, which could fall rapidly to lower levels. For the Milky Way and LSB dwarf galaxies, the replenishment hydrogen from the DM halo could be adequate to maintain their SFRs for much longer periods of time than for large, massive galaxies.

Regarding the paragraph from Chapter 23: Note that the initial high SFR of the large, massive galaxy is probably directly related to the amount the star-forming galaxy's disk overlaps the inner diameter of the DM halo's "hollow" core.

However, note in Chapter 56 that Dr. Mark Wilkinson of Cambridge University was quoted on 6 February 2006 as saying, "No matter what size, how bright, or how many stars they contained -- all the galaxies seemed to be sitting in roughly the same amount of dark matter."*50* Professor Gerry Gilmore of Cambridge made a similar statement at the time. This somewhat similar size of DM halos for "all the galaxies" could cause a hydrogen replenishment problem for the high SFR massive galaxies, leading to the early decline of their SFRs.

A plausible explanation has been provided for the mass-and-time-dependent SFR phenomena described by Cosmic DM Mystery #25. The explanation uses the same relativistic proton dark matter theory/cosmology used to explain the LSB dwarf galaxy star formation phenomenon, the Schmidt law, the starburst galaxy star formation phenomenon, and the formation of new blue stars in the spiral arms of galaxies. The author looks forward to the use of his theory/cosmology by others to solve Cosmic DM Mysteries #26, #27, #28, etc.

TABLE 3
RECAP

Decoding The Cosmos Via DM Relationism And By Solving The Cosmic DM Mysteries

PART II – Cosmic DM Mysteries #1 - #14

#1	Spiral Disk Galaxies Have Spherical Dark Matter Halos
#2	Accelerating Expansion Via Conserving DM Momentum
#3	Hydrogen Derived From DM Cosmic Ray Protons
#4	Magnetic Fields Derived From DM Cosmic Ray Protons
#5	Intersecting DM Filaments Create Galaxy Clusters
#6	Mature Galaxies Discovered In The Very Early Universe
#7	Dark Matter Spherical Cored Halos Have "Hollow" Cores
#8	Source Of Spiral Galaxies'/Halos' Angular Momentum
#9	No Central Dark Matter Cusp Found In Spiral Galaxies
#10	LSB Dwarf Galaxies Have Low Star Formation Rates
#11	LSB Galaxies Have Inclining Star Rotation Curves
#12	Galaxy Hydrogen Is Replenished From Halo Dark Matter
#13	Dark Matter, Hydrogen, Helium, And Muons Create Stars
#14	Earthbound Cosmic Ray Protons Depart From 4 Locations

PART III – Cosmic DM Mysteries #15 - #25

#15	Astrophysical Emergence Of Dark Matter Halos, After Eons
#16	UHECRs Arrive At Earth From Galaxy Superclusters
#17	Starburst Galaxies Form Via Merging Galaxy Clusters
#18	UHECR Protons Via Starburst Galaxies/Merging Galaxies
#19	Blue Stars In Spiral Arms Vs. Red Stars In Galaxy Nucleus
#20	Magnetic Field, DM Proton Energies Set Galaxy Halo Size
#21	Different Dark Matter For Small Galaxies And For Clusters
#22	800 Galaxies Detected, Less Than 1.2 Billion Years Old
#23	Fine Balance Between Dark And Baryonic Matter In Spirals
#24	Schmidt Law: SFR Vs. Surface Hydrogen Molecular Density
#25	Mass-And-Time-Dependent SFR Graphs For Field Galaxies

CHAPTER 55

SOME CONCLUSIONS AND CONSIDERATIONS

The relativistic proton dark matter theory and cosmology disclosed and explained in this book have been successful in providing plausible explanations for 25 cosmic mysteries regarding cosmological or astrophysical phenomena. Let us try to understand this theory/cosmology a bit better.

Considerable effort was made to avoid biasing the outcome of the Ockham's razor logic/relationism search for a dark matter candidate fitting the intrinsic dark matter description laid down by the Cosmic DM Mysteries and their related constraints. No dark-matter-linked Cosmic DM Mystery was knowingly excluded. To assure more confidence in the *dark matter relationism* selection process, the author also attempted to maximize the number of Cosmic DM Mysteries, and eventually utilized 25 of them.

As stated earlier, *dark matter relationism* is epitomized by the author's use of cosmic mysteries and *relationism* to tentatively identify dark matter and then to confirm its validity by using the same dark matter candidate to provide plausible explanations for additional cosmic mysteries, including some previously *not known to be related to dark matter*.

Until now, the cold dark matter theory has provided the most widely used theoretical dark matter candidates during the past 20 years -- WIMPs and neutralinos -- but unfortunately neither of these particles has been detected anywhere in the Universe using either astronomical or particle accelerator techniques. After 20 years, it is necessary that other dark matter candidates be given an equal opportunity to compete using equal standards of acceptance.

Furthermore, none of the following 15 astrophysical or cosmological phenomena seems to be explainable by the cold dark matter theory of WIMPs and neutralinos, whereas the relativistic proton dark matter theory seems to provide plausible explanations for all 15 of them.

Among The 25 Cosmic DM Mysteries, New And Plausible Explanations Are Provided In This Book For These 15 Well-Known Astrophysical Or Cosmological Phenomena That Had Not Been Adequately Explained Previously:

1. The "knees" and "ankle" of the cosmic ray proton energy distribution graph at the Earth.

2. The low SFRs of low surface brightness dwarf galaxies; the higher SFRs for spiral galaxies; and the much higher SFRs for large, massive spiral galaxies for limited periods of time.

3. The ignition of hydrogen fusion reactions in the first-generation stars.

4. An accelerating expansion of the Universe.

5. The source of magnetic fields in spiral galaxies.

6. How dust particles facilitate or expedite hydrogen fusion reactions in stars.

7. The empirical Schmidt law correlating star formation rate and the average molecular hydrogen density on the surfaces of isolated spiral galaxies.

8. Strongly self-interacting dark matter particles in halos of dwarf and LSB galaxies in contrast to weaker self-interacting dark matter particles in the halos of galaxy clusters, implying the existence of two types, forms, or modes of dark matter particles.

9. The extremely high star formation rates of starburst galaxies created by the merging of spiral galaxy clusters.

10. The blue and blue-white stars, as young as one million years old, in the spiral arms of mature spiral galaxies, which also contain five-billion-year-old red stars in their nuclei.

11. The process of two galaxy clusters merging that results in a source of ultra-high energy protons called ultra-high energy cosmic rays, or UHECRs.

12. The various departing locations of Earthbound cosmic ray protons with energy levels at or below 5×10^{19} eV.

13. That dark matter halos are almost spherically shaped, while their enclosed spiral galaxies are disk shaped.

14. That starburst galaxies usually exhibit new blue star formation primarily in their galaxy nuclei, while spiral galaxies exhibit new blue star formation in their spiral arms.

15. The growth of the spiral galaxy disk of the Andromeda galaxy (M31) by a factor of three, by the accretion of hydrogen.

These previously unexplained astrophysical or cosmological phenomena were selected because they are better known by astronomers, astrophysicists, and cosmologists.

Relativistic Proton Dark Matter May Be The Source Of The Stars, Planets, and DNA Changes:

Relativistic proton dark matter may be the source of sunlight, starlight, the stars, planets, and DNA changes throughout the Universe. Through this book, we have learned that proton dark matter probably feeds the hydrogen nuclear fuel to all the galaxies and that it ionizes atomic hydrogen, thereby forming hydrogen molecules faster. The proton dark matter also creates muons that catalyze hydrogen fusion nuclear reactions, and it bombards and triggers hydrogen-based muonic ions that ignite the stars. Through these four roles, the relativistic dark matter protons seem to have created all the stars and therefore all the planets in the Universe.

Few Answers From The Mainstream Star Formation Theory:

On the above list of 15 cosmic phenomena, the author believes that items 3, 6, 7, and 10 are surface-related star formation phenomena. These star formation phenomena cannot be explained by the generally accepted mainstream theory of star formation where clouds of hydrogen molecules collapse anywhere in a galaxy under their own weight and are heated through compression to hydrogen fusion temperatures. Whereas the relativistic proton dark matter theory appears to provide plausible explanations for all four of these star formation phenomena, the mainstream star formation theory fails to provide any explanations for them.

More Weaknesses In The Cold Dark Matter Theory?

There is no cold dark matter explanation for NASA's announcement on September 9, 2004,[33] that in addition to forming halos around galaxies and around galaxy clusters, dark matter is formed into long, large, dark matter filaments which create galaxy clusters where the dark matter filaments intersect. Not even an *astrophysical emergence* explanation would apply to CDM because the WIMP particles of the CDM theory are *weakly interacting.*

On the other hand, *astrophysical emergence* of DM filaments may be applicable to the relativistic DM proton because of its strong and various coulomb, electromagnetic, and nuclear interactions, as discussed in Chapter 44. However, more research on DM filaments is required to arrive at a definitive answer.

Furthermore, whereas NASA's report on the large, long, DM filaments forming galaxy clusters where they intersect implies a top-down theory of cluster formation, the CDM theory posits only a bottom-up theory of cluster formation.

There are over 100 astro-ph papers on the physics arXiv on the subject of *cored* dark matter halos/haloes, the formation of which is not explainable by the CDM theory but is easily explainable by the relativistic proton dark matter theory.

Do the computer simulation solutions arrived at by the CDM researchers represent the creation of the large-scale structure of the Universe formed by 100 billion uncharged theoretical WIMPs, or perhaps formed by 100 billion galaxies?

Why Another Dark Matter Approach Is Necessary:

Many cosmologists believe that cold dark matter seems to explain the development of the large-scale structure of the Universe better than does warm dark matter. They also believe that warm dark matter seems to explain the formation of galaxies better than does cold dark matter. This dilemma faced by the cosmologists today probably was created by three unnecessary constraints that pioneering cosmologists apparently placed on the dark matter particles; namely, that the dark matter particles have no coulomb charge, are not influenced by magnetic fields, and cannot be transformed into normal baryonic matter.

Drexler's relativistic proton dark matter theory/cosmology avoids these three "unnecessary" constraints and provides plausible explanations for the 15 well-known astrophysical or cosmological phenomena listed above. Drexler invites the proponents of cold and warm dark matter to offer their explanations for these 15 cosmic mysteries.

The author hopes that the use of relationism,. astrophysical emergence, and Ockham's razor logic can be applied successfully to other unexplained astrophysical and cosmological phenomena.

CHAPTER 56

EPILOGUE

The Local Group's Dwarf Spheroidal Satellite Galaxies Help Define Dark Matter

After a draft manuscript of this book had been completed, research astronomers at Cambridge University reported at a press conference on February 3, 2006 that their research on 10 dwarf spheroidal (dSph) satellite companions of the Milky Way and Andromeda revealed surprising information about dark matter. Researcher Professor Gerry Gilmore, according to the press, reported that:

1. Dark matter particles are not slow and cold, but instead appear to be moving at 9 kilometers per second and have an apparent temperature of about $10,000°$ C, which is higher than the $6,000°$ C at the surface of the sun.

2. "The strange thing about dark matter is that it doesn't give off radiation."[51]

3. "There must be some form of repulsion [between the dark matter particles] ...."[52] ... "We have to start looking into the physics of the interactions between dark

matter particles -- not just at the way they respond to gravity."⁵⁰

4. "This indicates that dark matter clumps together in building blocks which have a minimum size," said team member Dr. Mark Wilkinson in an article by Zeeya Merali in NewScientist.com on February 6, 2006, entitled, "'Tepid' temperature of dark matter revealed."⁵⁰ "This is 1,000 light years across, with 30 million times the mass of the sun," said Dr. Wilkinson.

In BBC NEWS, 6 February 2006, in an article entitled, "Dark matter comes out of the cold," Jonathan Amos reported that researchers have been able to establish that the galaxies [studied by Cambridge researchers] contain about 400 times the amount of dark matter as they do ordinary matter.

Dr. Wilkinson, a member of Professor Gilmore's team, was further quoted in the February 6, 2006 NewScientist.com article as saying, "No matter what size, how bright, or how many stars they contained -- all the galaxies seemed to be sitting in roughly the same amount of dark matter."⁵⁰ Does Wilkinson's statement imply that the dark matter halos existed prior to the formation of the enclosed galaxies?

In the same article, Dr. Robert Minchin, an astronomer at the Arecibo Observatory in Puerto Rico, is quoted as saying, "It [the results] contradicts the WMAP findings, which suggested that warm dark matter was unlikely."[50]

In the Education.Guardian, an article was published by science correspondent Alok Jha on February 6, 2006, entitled, "Research into dwarf galaxies starts to unlock the deep secrets of dark matter."[51] A bullet title reads, "1,000-light-year-wide bricks make up the universe."

In a February 10, 2006 article in the journal Science, entitled, "Dwarf Galaxies May Help Define Dark Matter,"[52] Daniel Clery reported that Gilmore suggests that they [the dark matter particles] interact with one another to spread out evenly. "There must .be some form of repulsion [between the dark matter particles] ...."

In National Geographic News, an article was published by James Owen on February 13, 2006, entitled, "Dark Matter properties 'Measured' for the First Time, Study Says," which quotes Professor Gilmore. Each galaxy studied was found to contain the same amount of dark matter. "That was a big surprise," Gilmore added. "The galaxies contain very different numbers of stars, so like everybody else we thought they would cover a very wide range of masses, but they don't. There seems to be a minimum mass for a galaxy." Do these statements by Gilmore imply that the dark matter halos existed prior to the formation of the enclosed galaxies?

On the morning of February 23, 2006, Professor Gilmore gave an invited progress report on his dark matter research by video conferencing at "Dark Matter 2006," the 7th UCLA Symposium on Sources and Detection of Dark Matter and Dark Energy in the Universe. The principal change from his previous data was that his recent estimate for the minimum dark matter mass is 50 million solar masses compared to 30 million solar masses estimated previously.

Since no research paper has been published on the Cambridge University findings, these phenomena have not been designated a Cosmic DM Mystery even though the Cambridge University three-year research program's results appear to be significant. The further confirmation of the following five different types of astronomical data could provide strong support for the Drexler dark matter theory/cosmology and could weaken support for the cold dark matter theory:

1. The measurements of the high-velocity/high-temperature dark matter particles.

2. The lack of thermal radiation from the 10,000° C dark matter particles.

3. Indications of strong self-interactions between dark matter particles.

4. The minimum size and mass of the dark matter clumps.

5. The contradiction of the earlier WMAP findings, which have provided support for the cold dark matter model.

Professor Gilmore's statement, "The strange thing about dark matter is that it has temperature but it does not give off radiation," raises a question that may be answered by the relativistic proton dark matter theory. The relativistic cosmic ray protons that enter the Earth's atmosphere are very fast and have very high energy but have low entropy and, therefore, would exhibit very little thermal radiation. Of course, when they decelerate, they radiate photons as when they enter a magnetic field or collide with atoms, molecules, photons, or dust, but only a portion of this radiation would be thermal radiation. Thus, Professor Gilmore's above statement essentially describes low-entropy relativistic dark matter protons and, therefore, his statement seems to provide support for the relativistic proton dark matter theory.

Professor Gilmore's statement, "We have to start looking into the physics of the interactions between dark matter particles -- not just at the way they respond to gravity," has been addressed in different ways in Chapter 44 and Chapter 50 of this book, and the mutual repulsion of positively charged protons is long known.

Chapter 44 is entitled, "Astrophysical Emergence Of Dark Matter Halos And Long, Large, Dark Matter Filaments Could Place Constraints On The Identity Of Dark Matter Particles." As pointed out in that chapter, the author believes that without the strong self-interaction of the dark matter particles, the dark matter halos around spiral galaxies and their galaxy clusters would not have formed under *astronomical emergence* or *emergent evolution* principles.

Chapter 50 is entitled, "Two Different Types Of Dark Matter Halo Particles Reported For Smaller Galaxies And For Galaxy Clusters." Chapter 50 relates that researchers have discovered that the self-interaction of dark matter particles orbiting galaxy clusters is much weaker than the self-interaction of dark matter particles orbiting LSB and dwarf galaxies. The relativistic proton dark matter theory is utilized to explain the difference.

Dr. Mark Wilkinson's (and Professor Gilmore's) statements that all dark matter halos around galaxies have roughly the same amount of dark matter, support Drexler's relativistic proton dark matter theory. This theory posits (see Jerome Drexler's April 2005 paper, astro-ph/0504512,[37] pages 12-13) that, "Dark matter halos around galaxies and galaxy clusters have outer diameters and 'hollow' core diameters determined by the galactic and extragalactic magnetic fields and the energy spectrum of the relativistic protons." The Drexler paper further states, "Also, the author believes that the outer diameter and core size diameter of DM halos are not significantly affected by the amount of galaxy mass enclosed."

The discovery by the Cambridge researchers that dark matter is found in minimum-size building blocks of 1,000 light-years across is consistent with the earlier discovery of the diminutive galaxy I Zwicky 18 (see Chapter 39), which is just 3,000 light-years across.

One possible explanation for the minimum size dark matter building blocks of 1,000 light-years across will now be presented in the following five paragraphs, based on Drexler's relativistic proton dark matter theory:

Synchrotron radiation (or emission) is electromagnetic radiation that is emitted by charged particles moving at relativistic velocities in circular/spiral orbits in an orthogonal magnetic field. The rate of synchrotron emission is inversely proportional to the product of the radius of curvature of the particle's orbit and the fourth power of the mass of the particles. The radius of curvature of the particle orbits is determined by the Larmor Radius equation. The Larmor Radius is directly proportional to the kinetic energy of the particles and inversely proportional to the magnetic field orthogonal to the paths of the particles.

The 10 dwarf spheroidal satellite galaxies of the Milky Way/Andromeda studied by Gilmore and his group all lie relatively close by their host galaxy. For example, the

Fornax spheroidal galaxy is approximately 138 kpc from the Milky Way. (Dwarf irregular galaxies lie much further away from their host galaxies.)

Thus, the two magnetic fields that influence the radius of curvature of the proton paths in the dark matter halos of a dwarf spheroidal satellite galaxy are the magnetic field strength of its host galaxy (2×10^{-6} gauss for the Milky Way) and the much lower extragalactic (intergalactic) magnetic field strength (estimated at 1×10^{-9} gauss).

Therefore, in the vicinity of the DM halo of the dwarf spheroidal galaxy (dSph) 138 kpc from the Milky Way, the combined magnetic field strength probably would be above 1×10^{-8} and less than 1×10^{-7} gauss. This magnetic field strength may indicate that the DM halo of the dSph would be more than one order of magnitude smaller in diameter than that of the Milky Way's DM halo. Also, the average synchrotron radiation losses of the dSph's DM halo's protons may be more than one order of magnitude greater than the average synchrotron radiation losses for protons in the DM halo of the Milky Way.

These synchrotron radiation losses of the halo protons of the dSph galaxies, in turn, would lower the average kinetic energy of the orbiting dark matter protons, which would further lower the orbiting protons' radius of curvature. This, in turn, would further increase the synchrotron radiation energy losses of the orbiting protons. These energy losses eventually could accelerate and destroy the stability of the protons' orbital paths, thereby providing some evidence that relativistic proton dark matter halos of the dSph galaxies may not be able to exist stably below some minimum size radius and related minimum mass.

The newly discovered characteristics of the dark matter particles described by the Cambridge University researchers and mentioned in this chapter appear to have plausible explanations based upon the relativistic proton dark matter theory/cosmology.

The author's December 2003 book probably represents the first publication in more than 10 years positing that dark matter particles may be moving at a relatively high velocity. The author's 19-page follow-up paper posted April 22, 2005 on the Physics arXiv as astro-ph/0504512[37] appears to be the second such publication.

The anticipated Cambridge University research paper could have the necessary astronomical data to evolve as an important dark matter paper.

While waiting for the relevant Cambridge research paper to be published, it may be worthwhile for interested scientists to read three 2006 research papers about dwarf spheroidal galaxies in the Local Group. Two are by the Gilmore/Wilkinson Cambridge University team. They are astro-ph/0511759 v2[53] dated 6 Jan 2006 and astro-ph/0602186 v1[54] dated 8 Feb 2006. The third paper, by a U.S.-based team, was posted on 29 March 2006 as astro-ph/0603775v1 and is entitled, "A Large Dark Matter Core In The Fornax Dwarf Spheroidal Galaxy?"[55]

ACKNOWLEDGMENTS

I thank Sandra Faber, George Blumenthal, Kim Griest, and Alan Watson for their constructive critique of my dark matter theory or of my December 2003 astrophysics/cosmology book, *"How Dark Matter Created Dark Energy And The Sun."*

I thank Sir Martin Rees, Astronomer Royal and President of the Royal Society of the United Kingdom, for including my 2003 astrophysics/cosmology book among the five books recommended for reading in the field of cosmology in conjunction with his December 2004 TV science program in the United Kingdom, "What We Still Don't Know."

I thank the British Broadcasting Corporation's BBC Radio 4, for an early 2005 program feature called, "In Our Time -- Subject Research -- Dark Energy," which recommended seven books for further reading, including my 2003 astrophysics/cosmology book.

APPENDIX A

The Scientific Community's Long-Held Objections To Any Proton Dark Matter Theory Are Summarized By The Author As Follows:

1. Scientists' arguments regarding nucleosynthesis, abundance ratios, and the amount of deuterium are as follows: Deuterium was created only in the very early Universe. Since deuterium is fragile, it is not present in the stars. The more deuterium there is today in the Universe, the fewer baryons there are since baryons convert deuterium. The high abundance of deuterium in the Universe today implies a density of ordinary baryonic matter of between 5% and 10% of the critical mass density of the Universe. The critical mass density is the minimum mass density required if the expansion of the Universe is eventually to cease. Since the scientific community believes that the actual mass density of the Universe is much greater than the 5% to 10% of the critical mass density, scientists believe that dark matter must be something other than baryons.

2. The measured cosmic microwave background (CMB) fluctuations indicate that the mass density ripples were too small in the early Universe to attract baryons and evolve into galaxies. To attract baryons before the mass density ripples would disperse, the CMB fluctuations must be of the order of one part in 100, whereas mass density fluctuations observed by COBE are of the order of one part in 100,000. That also led the astrophysicists

and cosmologists to conclude that the vast majority of the matter in the Universe must be comprised of something other than baryons.

3. Regarding flatness (Euclidean geometry) and the critical mass density, scientists believe that the geometry of the Universe is closer to the ordinary Euclidean geometry than any other geometry. Also, it is known that if the Universe had the critical mass density, the Universe could be described in terms of Euclidean geometry and would therefore be "flat." Scientists believe that if all the luminous (ordinary) matter in the galaxies were dispersed in the Universe, the mass density of the Universe would be many times smaller than the critical mass density. From the speeds of stars around the centers of galaxies and the speeds of galaxies moving within galaxy clusters, it appears that dark matter in the Universe may be about 10 times more massive than luminous ordinary matter and about six times more massive than the total ordinary matter.

Author's Response Regarding The Above Three Objections To The Proton-Dark-Matter Candidacy:

For many years, baryons have been ruled out as a DM candidate because the primordial nucleosynthesis calculations and other cosmological considerations indicate a very low *particle abundance* of baryons in the Universe. However, this argument does not address the large quantity of relativistic protons/baryons in the Universe (see Chapter #24). For example, a 15% ratio of relativistic protons to non-relativistic baryons in the Universe satisfies the particle-abundance-maximum constraint and also satisfies the mass-abundance-minimum constraint because the DM particles, having a very large relativistic mass, are still able to form 80% to 85% of the mass of the Universe.

It is widely accepted by cosmologists that ordinary matter in the Universe totals about 4% of the total mass/energy and that dark matter totals about 23% of the total mass/energy; therefore, the mass energy of dark matter in the Universe is about six times higher than that of ordinary matter.

This same ratio appears plausible for relativistic proton dark matter. If, for example, the total number of relativistic dark matter protons in the Universe were no more than 15% of the total number of baryons in the Universe, the relativistic proton dark matter theory would be compatible with the proton limitations of each of the above three objections of the scientific community.

That is, if the average relativistic mass of the dark matter protons is, say, 50 times the rest mass of a proton and their number is, say, about 12% of the number of baryons in the Universe, then the dark matter relativistic mass would total about six times that of ordinary matter mass in the Universe. This matches cosmologists' estimates mentioned above. For an average relativistic mass ratio of 50, the average proton energy would be about 5×10^{10} eV, which is near the low end of the cosmic ray proton energy range of about 1×10^{10} eV to 5×10^{19} eV (see Appendix B, Slide #17).

The Objection Of Some Scientists Is Based On The GZK (Greisen-Zatsepin-Kuzmin) Cut-Off Theory Argument:

In the past, the 1966 GZK cut-off theory was sometimes used as one of the arguments to counter the relativistic proton dark matter theory. The GZK cut-off theory predicts that because of the interaction with the cosmic microwave background, relativistic protons cannot have energies higher than 6×10^{19} eV at the Earth, since above those levels they would lose energy rapidly in collisions with the CMB.

Author's Response To The GZK Argument:

Some astrophysicists who have been inclined to use the 1966 GZK cutoff theory to rule out the relativistic proton dark matter model were apparently not aware of the 1998 paper designated hep-ph/9808446 and entitled, "Evading the GZK

Cosmic-ray Cutoff,"[56] by Sidney Coleman and Nobelist Sheldon L. Glashow, or the 1997 published chapter of Nobelist James W. Cronin in *Unsolved Problems in Astrophysics*.[57] James Cronin's chapter in the book reported two observed cosmic ray protons with energies about four times higher than the theoretical GZK proton-energy cutoff. The presence of UHE cosmic ray protons in the solar system, including a very few with energies well above the GZK cutoff, adds some plausibility to an anti-GZK effect.

Also, researchers have reported an anti-GZK effect that arises when UHE relativistic protons moving through an intergalactic magnetic field experience diffusive propagation. They report that this effect causes a jump-like increase in the distance UHE protons can travel. From its description in the literature, apparently relativistic protons orbiting a galaxy in its halo's magnetic field also would experience an anti-GZK effect. For example, see astro-ph/0507325 authored by R. Aloisio and V. S. Berezinsky.[58]

Further, it should be noted that under the relativistic proton dark matter theory/cosmology, the only relativistic protons that could have energies even close to the theoretical GZK cutoff are those orbiting a galaxy supercluster. For this case, the proton flux density would be extremely low (see Appendix B, Slide #17). For example, cosmic ray protons at an energy level of 10^{19} eV or above striking the Earth's atmosphere total only 3 to 4 per square kilometer per century.

APPENDIX B

Presented here are 18 selected pages involving six important references from J. Drexler's book, *"How Dark Matter Created Dark Energy And The Sun"* (Universal Publishers Boca Raton, Florida USA, 2003).

Energies of Relativistic Protons Versus Their Relativistic Mass -- Where Do They Exist in Nature?

Energy 9.38 x 10^8 eV	Relativistic Mass of a Proton In Terms of Its Rest Mass, m_o m_o
10^{10} eV	11 m_o
10^{11} eV	110 m_o
10^{12} eV	1,100 m_o
10^{13} eV	11,000 m_o
10^{14} eV	110,000 m_o
10^{15} eV	1,100,000 m_o
10^{16} eV	11,000,000 m_o
10^{17} eV	110,000,000 m_o
10^{18} eV	1,100,000,000 m_o

$$\text{Relativistic Mass} = \frac{\text{Energy (in joules)}}{C^2 \text{ (in meters/sec)}}$$

Page 22 of *How Dark Matter Created Dark Energy And The Sun*

SLIDE #15

Such Highly Energetic Protons Can Be Found Striking The Earth's Atmosphere As Cosmic Ray Protons

Approximate Kinetic Energy	Approximate Cosmic Ray Flux On the Earth's Atmosphere
10^8 to 10^{10} eV	Slightly less than 1,000 particles per square meter per second
10^{11} eV	One particle per square meter per second
7×10^{15} eV	One particle per square meter per year
3×10^{18} eV	One particle per square kilometer per year
10^{19} eV	3 to 4 particles per square kilometer per century

Page 23 of *How Dark Matter Created Dark Energy And The Sun*

SLIDE #16

Cosmic-Ray Energy Distribution At The Earth*

CERN Courier, Vol. 35, No. 10, December 1999

[See Slide #16 for the Key Data Points]

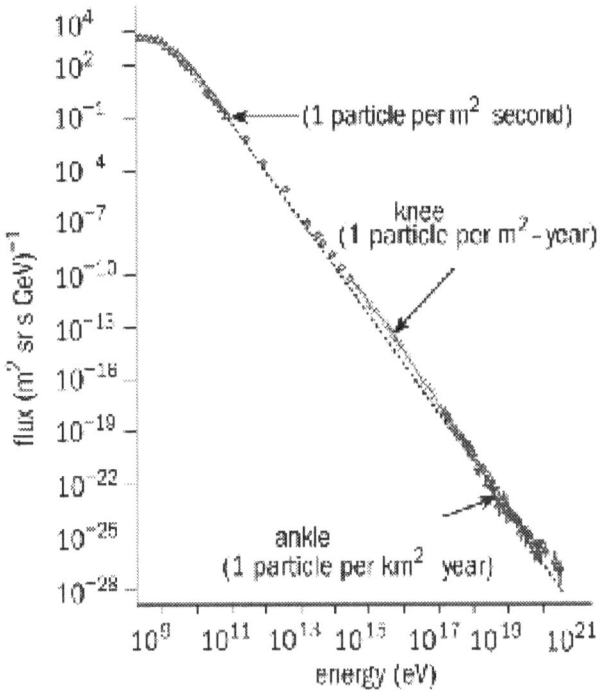

"The cosmic-ray energy distribution shows remarkable uniformity over 10 orders of magnitude. However, there are two kinds. The ACCESS experiment is designed to investigate 'the knee' (near 10^{15} eV)."

*Available on the Internet at http://www.cerncourier.com/main/article/39/10/8/1.

Page 24 of *How Dark Matter Created Dark Energy And The Sun*

SLIDE#17

The Accelerating Expansion Between Galaxy Clusters (Not Between Galaxies)

- Note that Michael Turner points out (see slide #54) that the dark energy "isn't found in galaxies or even clusters of galaxies." This is an extremely important point. This statement means that dark energy pushes galaxy clusters apart but doesn't push galaxies apart within clusters and doesn't push stars apart within a galaxy.

- This is observed locally. While the Universe is expanding, in our Local Group of galaxies Milky Way and Andromeda are moving towards each other at 119 km/sec. Also, our Local Group is moving toward the Local Supercluster at 600km/sec. This illustrates that local strong gravitational forces can overcome the dark energy antigravity forces. This also suggests that perhaps in the earlier, smaller, and denser Universe, when the Universe was less than about 8.2 billion years old, the closeness of the galaxy clusters favored dark matter gravity over dark energy antigravity, and expansion acceleration could not take hold. When the Universe was 8.7 billion years old, the expansion acceleration began, according to Adam Riess. (See slide #53.)

Page 65 of *How Dark Matter Created Dark Energy And The Sun*

Jerome Drexler's Theory Of The Accelerating Expansion Between Galaxy Clusters

- As the synchrotron radiation emission of gamma rays continues, not only does the kinetic energy of the halo UHE protons fall, but their relativistic mass will fall as well (slides #39 and #40). See slide #15 and assume that the average kinetic energy of the UHE protons declined from 10^{16} eV to 5×10^{15} eV over a period of time in the dark matter halo of some galaxy cluster. This represents a decline in the dark matter mass in the halo of that galaxy cluster of 50% and perhaps a 40% decline in the mass of the combined galaxy cluster and its halo.

- Such a reduction in the dark matter halo mass around galaxy dusters would:

    (1) Raise the galaxy clusters' velocities under the Law of Conservation of Linear Momentum. (See footnote on slide #75.) [Drexler]

    and

    (2) Reduce each galaxy cluster's gravitational attraction to nearby galaxy clusters, thereby facilitating their more rapid separation. [Drexler]

Page 66 of *How Dark Matter Created Dark Energy And The Sun*

SLIDE #59

Drexler's Theory Of The Accelerating Expansion Between Galaxy Clusters

- The gravitational attraction effect of item (2) on the previous slide will diminish through the years as the nearby galaxy clusters become more distant. Note from slide #54 that from the age of 300,000 years until today, the spacings between galaxies increased by a factor of 1,000.

- In an expanding Universe, all galaxy clusters are moving away from each other. Meanwhile, the masses of their dark matter halos of UHE protons are declining because of the synchrotron radiation energy losses. As a result, the velocity of every galaxy cluster should rise (owing to the reduced gravitational attraction between them and the Law of Conservation of Linear Momentum), thereby accelerating the expansion of the Universe. [Drexler]

Page 67 of *How Dark Matter Created Dark Energy And The Sun*

Drexler's Theory Of The Accelerating Expansion Between Galaxy Clusters

- The antigravity repulsion between galaxy clusters is proportional to the relative decrease in mass of their dark matter halos of UHE protons owing to the protons' loss of kinetic energy through synchrotron radiation. [Drexler]

- Gravitational attraction is directly proportional to the product of the masses of two galaxy clusters and inversely proportional to the square of the distance between them.

- Thus, for the earlier, smaller, and denser Universe more than five billion years ago, the smaller distances between galaxy clusters caused the gravitational attraction between them to be high because of the inverse square of those smaller distances between galaxy clusters. This inverse-square relationship may be a principal reason that the accelerated expansion did not begin until five billion years ago when the conservation-of-momentum effect and reduced galaxy cluster mass finally overcame the gravitational attraction between galaxy clusters. [Drexler]

Page 68 of *How Dark Matter Created Dark Energy And The Sun*

SLIDE #61

Drexler's Theory:
UHE Cosmic-Ray Nuclei May Have Facilitated The Triggering Of The Sun's Fusion Reaction

- With the UHE cosmic ray protons having a kinetic energy in the range of 10^{10} to 10^{20} electron volts, they not only added mass and fuel to the formation of the Sun but considerable nucleus-to-nucleus collision energy as well.

- These UHE nuclei provide a clue that the UHE protons and heavier high-collision-cross-section UHE nuclei may have facilitated the triggering of the Sun's fusion reactions and its birth.

- The traditional theory of the Sun formation involving a hydrogen gas cloud, forces of gravity, compression, and high temperature heating may have to be modified. (For further information about the UHE cosmic-ray nuclei, see slide #17, an energy distribution graph that is a plot of cosmic-ray particle flux versus particle energy.) [Drexler]

What Is The Difference Between A UHE Proton And A Cosmic-Ray Proton Bombarding the Earth?

- When a UHE proton in the halo of the Milky Way (a spiral galaxy) loses a significant portion of its kinetic energy over billions of years through synchrotron radiation, the proton will eventually plummet into the galaxy, thereby accelerating its energy loss. It then becomes one of the cosmic ray protons which bombard wide regions of the galaxy, including the solar system, and may have played a role in creating the Sun and other stars as explained in slides #62 – #71.

- With UHE protons remaining active for billions of years, they may be thought of as "immortal" UHE protons while at end of their life they become "mortal" cosmic ray protons plummeting into the galaxy in what I call a "death spiral."

Page 85 of *How Dark Matter Created Dark Energy And The Sun*

SLIDE #78

The Transformation Of "Immortal" UHE Protons Into "Mortal" Cosmic-Ray Protons Through The "Death Spiral"

- The synchrotron radiation loss of a relativistic charged particle is inversely proportional to both the radius of curvature of its path and the fourth power of its mass.

- The radius of curvature of a UHE proton's spiral path is equal to the Larmor Radius (see slide #85) and is directly proportional to the kinetic energy of the proton and inversely proportional to the magnetic field strength.

- The extragalactic magnetic field is reported to be about 1×10^{-9} gauss while the magnetic field in the interior of the Milky Way is about 2,000 times greater, at 2×10^{-6} gauss.

- Therefore, in the extragalactic dark matter halo of the galaxy, the magnetic field is very weak, the kinetic energy of the protons is high, the synchrotron radiation losses are extremely low, and the UHE proton may be able to circulate for billions of years. [Drexler]

Page 86 of *How Dark Matter Created Dark Energy And The Sun*

SLIDE #79

The Transformation Of "Immortal" UHE Protons Into "Mortal" Cosmic-Ray Protons Through The "Death Spiral"

- After billions of years in the extragalactic halo of a spiral galaxy, some of the UHE protons should eventually lose enough energy that their spiral paths will be reduced in diameter and the UHE protons will approach the surface of the galaxy. When the UHE halo protons enter the galaxy, their energy has perhaps diminished by a factor of 10 or so and the magnetic field might be about 100 times greater, thereby increasing the synchrotron radiation loss by a factor of 1,000. (The radius of a proton's spiral path is called the proton's Larmor Radius, which can be calculated as shown in slide #85.)

- Thus, as a UHE proton enters the galaxy, its energy will plummet further, say to half, thereby doubling the synchrotron radiation loss. The proton's energy will drop more and more, and it will enter into the "death spiral" as it plunges deeper into the galaxy and begins to be known as a cosmic-ray proton.

Page 87 of *How Dark Matter Created Dark Energy And The Sun*

The Transformation Of "Immortal" UHE Protons Into "Mortal" Cosmic-Ray Protons Through The "Death Spiral"

- The rate of synchrotron radiation is inversely proportional to the fourth power of the mass of the particle. Therefore, synchrotron radiation from protons is infinitesimal compared to synchrotron radiation from electrons. More precisely, proton synchrotron radiation losses are lower than radiation losses from electrons, following the same radius of curvature path, by a factor of about 11 trillion (the proton/electron mass ratio of 1,836 to the fourth power).

- The discussion on slides #79 and #80 has to do with spiral galaxies, which are known to have a dark matter halo and also a black hole containing a few million solar masses.

Drexler's Cosmic-Ray Cosmology Applied To Galaxy Formation

The Proton Larmor Radius

- A proton crossing an orthogonal magnetic field enters into a spiral path. The radius of one cycle of that spiral path is called the Larmor Radius.

- The Larmor Radius of a proton can be calculated as:

$$r = 110 \text{ Kpc} \times \frac{10^{-8} \text{ gauss}}{B} \times \frac{E}{10^{18} \text{ eV}}$$

  where Kpc means kilo parsec and
  one parsec equals 3.26 light-years

- The galactic magnetic field within the Milky Way is approximately 2×10^{-6} gauss compared to the extragalactic magnetic field at 1×10^{-9} gauss.

Page 92 of *How Dark Matter Created Dark Energy And The Sun*

Drexler's Cosmic-Ray Cosmology Applied To Galaxy Formation

The Proton Larmor Radius

- The Larmor Radius for a 10^{16} eV proton in the Milky Way halo's extragalactic magnetic field of 10^{-9} gauss is 11 Kpc; for a 10^{17} eV proton it is 110 Kpc; and for a 10^{18} eV proton it is 1,100 Kpc.

- The diameter of the Milky Way galaxy is about 100,000 light-years or 30.7 Kpc and its radius is about 15 Kpc.

- Studies in 1999 found that the dark matter halo of a spiral galaxy extends about 10 to 20 times the size of the visible regions (slide #3). Using a factor of 15, the radius of the dark matter halo would extend to perhaps 225 Kpc.

- Thus, some 10^{16} eV protons with a Larmor Radius of 11 Kpc might stay within the Milky Way galaxy's 15 Kpc radius. A 10^{17} eV proton with a Larmor Radius of 110 Kpc might remain within the halo's 225 Kpc radius, but a 10^{18} eV proton would probably leave both the galaxy and the halo.

Page 93 of *How Dark Matter Created Dark Energy And The Sun*

Only One Dark Matter Candidate Establishes The Approximate Size Of The Milky Way

- From slides #85 and #86 it has been shown that the Larmor Radius for a 10^{16} eV proton in the Milky Way's extragalactic magnetic field of 10^{-9} gauss is 11 Kpc; for a 10^{17} eV proton it is 110 Kpc; and for a 10^{18} eV proton it is 1,100 Kpc.

- In slide #17 entitled, "Cosmic-Ray Energy Distribution for the Milky Way," the "knee" of the curve falls on a proton energy of approximately 5×10^{15} eV, which means that some protons with that energy might stay within the galaxy.

- From the above it could be concluded that the Milky Way galaxy should have a minimum radius of about 11 Kpc, compared to astronomers' estimate of about 15 Kpc.

- This seems to provide additional evidence that UHE protons are a credible dark matter candidate. No other currently proposed dark matter candidate can be used to estimate the size or even the order of magnitude of the size of the Milky Way galaxy.

Page 94 of *How Dark Matter Created Dark Energy And The Sun*

SLIDE #87

Drexler's Cosmic-Ray Cosmology Applied To Galaxy Formation

Some Plausible Speculations

- When a UHE proton moving along a certain path crosses orthogonal magnetic field lines, it is deflected up or down depending upon the magnetic field direction. The degree of deflection is proportional to the orthogonal magnetic field strength.

- Any magnetic deflection of a UHE proton reduces its velocity in the direction of its original path for two reasons: the deflection itself will cause its direction to change, and synchrotron radiation losses will reduce its forward velocity.

- Thus, if UHE protons that are moving through space at a certain velocity encounter a bulge (increase) in the magnetic field strength, they will not pass through the magnetic bulge region as quickly as through a no-bulge region and could possibly linger in the region. To describe this magnetic attraction effect for relativistic protons and electrons, I will use the term "attract" in quotes.

Page 95 of *How Dark Matter Created Dark Energy And The Sun*

Drexler's Cosmic-Ray Cosmology Applied To Galaxy Formation

Some Plausible Speculations

- Since electrons lose energy through synchrotron radiation at a rate 11 trillion times faster than protons, electrons would lose their energy quickly and tend to circulate and accumulate in magnetic-bulge regions.

- Such electron-filled regions would add a coulomb attractive force to UHE protons, slowing them down and also facilitating their conversion into hydrogen atoms. Conceivably, this process could lead, in some cases, to negatively charged proto-galaxies or galaxies surrounded by positively-charged UHE proton dark matter. This, in turn, could lead to lightning-like proton electrical discharges creating a multitude of gamma-ray bursts (GRBs) of immense proportions.

- As the UHE protons slow down in close proximity to a magnetic-field bulge they could linger close by, adding mass and gravitational attraction to the magnetic bulge region.

Drexler's Cosmic-Ray Cosmology Applied To Galaxy Formation

Some Plausible Speculations

- Magnetic-field bulges "attract" relativistic electrons and protons. The relativistic proton and electron "attraction" to a magnetic bulge may be different from, but as real as, the gravitational attraction between two masses.

- Previously, my concept of the "death spiral" was discussed with regard to UHE protons plummeting into the galaxy from the halo. In that case the galaxy magnetic field was about 2,000 times the strength of the extragalactic field. The Milky Way is an example of a significant magnetic-field bulge covering a large area in space that should capture almost all UHE protons spiraling through. In the solar system, the Sun is an example of a very significant magnetic-field bulge. With its magnetic field strength of about 50 gauss and the Milky Way's magnetic field strength at 2×10^{-6} gauss, the magnetic field ratio is 25 million.

Page 97 of *How Dark Matter Created Dark Energy And The Sun*

SLIDE #90

REFERENCES

1. G. Blumenthal, S. Faber, J. R. Primack and M. J. Rees, 1984 *Nature* 311, 517

2. F. Zwicky, 1937 *Astrophys. J.* (Lett) 86, 217

3. V. C. Rubin, N. Thonnard and W. K. Ford, 1978 *Astrophys J.* (Lett) 225, L107

4. V. Rubin, *Bright Galaxies -- Dark Matters* (Amer. Inst. Physics, New York, 1997), p. 109-116

5. J. Drexler, *How Dark Matter Created Dark Energy And The Sun* (Universal Publishers, Parkland, Florida, USA, 2003), p. 22-24

6. J. Drexler, *op. cit.*, p. 92-94

7. A. M. Hillas, 1984, *Ann. Rev. Astron. Astrophys.*, 22, 425

8. A. A. Watson, 1985, 19th Intl. cosmic ray conference, La Jolla, USA, vol. 9, p. 111

9. G. Rudiger and R. Hollerbach, *The Magnetic Universe -- Geophysical and Astrophysical Dynamo Theory* (Wiley-VCH Verlag GmbH & Co. KGaA, Weinheim, Germany, 2004), p. 215

10. H. Lesch and M. Hanasz, 2003, *A&A* 401, 809-816

11. M. Hanasz, G. Kowal, K. Otmianowska-Mazur, and H. Lesch, 2004, astro-ph/0402662 v1

12. B. M. Gaensler et al, 2005, astro-ph/0503226 v1

13. B. M. Gaensler et al, 2005, astro-ph/0503371 v1

14. J. Drexler, *op. cit.* p.85-88

15. J. Drexler, *op. cit.*, p. 95-97

16. J. Drexler, *op. cit.*, p. 78

17. J. Bailin, C. Power, B. K. Gibson & M. Steinmetz, 2005, astro-ph/0502231 v1

18. S. Yoshida, *Ultra-high Energy Particle Astrophysics* (Nova Science Publishers, Inc. USA, 2003)

19. J. Drexler, *op. cit.*, p. 65-68

20. K. Glazebrook et al, 2004, Nature, **430**, 181

21. A. Cimatti et al, 2004, Nature, **430**, 184

22. P. J. McCarthy, D. L. Borgne, D. Crampton, H-W Chen, R. G. Abraham, K. Glazebrook et al, 2004, astro-ph/0408367 v1

23. R. Minchin et al, 2005, astro-ph/0502312

24. Y. I. Izotov, T. X. Thuan, 2004, astro-ph/0408391

25. P. Sokolsky, *Introduction to Ultrahigh Energy Cosmic Ray Physics* (Westview Press, member of the Perseus Books Group, Boulder Colorado, USA, 2004) p. 7-8

26. G. Cresci, L.Vanzi, and M. Sauvage, 2004, astro-ph/0411486 v2

27. A. S. Bishop, *Project Sherwood -- The U.S. Program in Controlled Fusion* (Addison-Wesley Publishing Company, Inc., Reading, Massachusetts, U.S.A. 1958) p 177-178

28. R. Clay and B. Dawson, Cosmic Bullets -- *High Energy Particles in Astrophysics* (Helix Books, Addison-Wesley, Australia, 1998)

29. A. A. Watson, 2002, *Contemporary Physics*, volume 43, number 3, p. 181-195

30. J. W. Cronin, 2004, astro-ph/0402487 v1

31. A. A Watson, 2004, astro-ph/0410514 v1

32. M. J. Rees, 2004, astro-ph/0402045 v1

33. "Motions in nearby galaxy cluster reveal presence of hidden superstructure," NASA Marshall Space Flight Center Release No. 04-231, Huntsville, Alabama, September 9, 2004, p. 1

34. C. A. Scharf, D. R. Zurek, and M. Bureau, 2004, astro-ph/0406216

35. "Astronomical surprise: Massive old galaxies starve to death in the infant universe," Carnegie Observatories Release No.050310, Pasadena, California, March 10, 2005

36. Ivo Labbe et al, 2005, astro-ph/0504219 v1

37. Jerome Drexler, 2005, astro-ph/0504512 v1

38. Elena Pierpaoli and Glennys Farrar, 2005, astro-ph/0507679 v3

39. "Galaxies and the Universe -- Starburst Galaxies," University of Alabama website:
http://www.astr.ua.edu/keel/galaxies/starburst.html

40. Diego F. Torres and Luis A. Anchordoqui, 2005, astro-ph/0505283 v1

41. R. Ibata et al, 2005, astro-ph/0504164 v1

42. Bo Qin, Ue-Li Pen, and Joseph Silk, 2005, astro-ph/0508572 v1

43. David N. Spergel and Paul J. Steinhardt, 1999, astro-ph/9909386 v2

44. "Young Galaxies Grow Up Together in a Nest of Dark Matter," Subaru Telescope/National Astronomical Observatory of Japan, press release, December 21, 2005.
http://subarutelescope.org/Pressrelease/2005/12/21/index.html

45. Stacy S. McGaugh, 2005, astro-ph/0509305 v1

46. Andrey V. Kravtsov, 2003, astro-ph/0303240 v1

47. Yuexing Li, Mordecai-Mark Mac Low, and Ralf S. Klessen, 2005, astro-ph/0508054 v1

48. Kevin Bundy et al, 2005, astro-ph/ 0512465 v1

49. "Large survey of galaxies yields new findings on star formation," University of California at Santa Cruz press release, January 9, 2006
   http://www.ucsc.edu/news_events/press_releases/text.asp?pid=797

50. "'Tepid' temperature of dark matter revealed," NewScientist.com news service, 6 February 2006
   http://www.newscientistspace.com/article.ns?id=dn8685&print=true

51. "Research into dwarf galaxies starts to unlock the deep secrets of dark matter," Guardian, 6 February 2006
   http://education.guardian.co.uk/print/0,,5392356-108229,00.html

52. "Dwarf Galaxies May Help Define Dark Matter" Science, 10 February 2006
   http://www.sciencemag.org/cgi/content/full/311/5762/758b

53. J. I. Read, M. I. Wilkinson, et al, 2006 astro-ph/0511759 v2

54. Mark I. Wilkinson et al, 2006, astro-ph/0602186 v1

55. Louis E. Strigari et al, 2006, astro-ph/0603775 v1

56. Sidney Coleman and Sheldon L. Glashow, 1998 hep-ph/9808446

57. J. N. Bahcall and J. P. Ostriker, *Unsolved Problems in Astrophysics* (Princeton Univ. Press, Princeton, New Jersey, 1997). Chapter 17 The Highest Energy Cosmic Rays

58. R. Aloisio and V. S. Berezinsky, 2005, astro-ph/0507325 v1

BIBLIOGRAPHY AND SUGGESTED SOURCES

The following is a list of books and articles that the author suggests as recommended reading. Readers should not assume that any author listed below or throughout this book agrees with the author's views or theories.

J. N. Bahcall and J. P. Ostriker, *Unsolved Problems in Astrophysics* (Princeton Univ. Press, Princeton, New Jersey, 1997).

A. S. Bishop, *Project Sherwood -- The U.S. Program in Controlled Fusion* (Addison-Wesley Publishing Company, Inc., Reading, Massachusetts, U.S.A. 1958).

R. Clay and B. Dawson, *Cosmic Bullets – High Energy Particles in Astrophysics* (Helix Books, Addison-Wesley, Australia, 1998).

K. Croswell, *The Alchemy of the Heavens* (Anchor Books, Doubleday, New York, 1995).

J. Drexler, *How Dark Matter Created Dark Energy And The Sun* (Universal Publishers, Parkland, Florida, USA, 2003).

D. Filkin, *Stephen Hawking's Universe* (Basic Books, New York, 1997).

M. W. Friedlander, *Cosmic Rays* (Harvard Univ. Press, Cambridge, MA and London, England, 1989).

M. W. Friedlander, *A Thin Cosmic Rain — Particles From Outer Space* (Harvard Univ. Press, Cambridge, MA and London, England, 2000).

H. Friedman, *The Astronomer's Universe* (W.W. Norton & Company, New York, 1998).

T.K. Gaisser, *Cosmic Rays and Particle Physics* (Cambridge University Press, Cambridge U.K.,1990).

D. Goldsmith, *The Astronomers* (St. Martin's Press, New York, 1991).

A. H. Guth, *The Inflationary Universe* (Helix Books, Perseus Books, Reading, Massachusetts, 1998).

R. P. Kirshner, *The Extravagant Universe -- Exploding Stars, Dark Energy and the Accelerating Cosmos* (Princeton Univ. Press, Princeton, New Jersey, 2002).

E. W. Kolb and M. S. Turner, *The Early Universe* (Addison-Wesley, USA, 1990).

L. Krauss, *Quintessence* (Basic Books, A Member of the Perseus Books Group, New York, N.Y., 2000).

T.S. Kuhn, *The Structure of Scientific Revolutions* (The University of Chicago Press, Chicago Illinois, 1970).

R. B. Laughlin, A Different Universe (Basic Books, A Member of the Perseus Books Group, New York, N.Y., 2005).

M. S. Longair, *High Energy Astrophysics,* Volume I, Second Edition (Cambridge Univ. Press, Cambridge, UK 1992).

M. S. Longair, *High Energy Astrophysics,* Volume II, Second Edition (Cambridge Univ. Press, Cambridge, UK, 1994).

M. S. Madsen, *The Dynamic Cosmos* (Chapman & Hall, London, England, 1995).

V. Rubin, *Bright Galaxies -- Dark Matters* (Amer. Inst. Physics, New York, 1997).

G. Rudiger and R. Hollerbach, *The Magnetic Universe -- Geophysical and Astrophysical Dynamo Theory* (Wiley-VCH Verlag GmbH & Co. KGaA, Weinheim, Germany, 2004).

D. W. Sciama, *Modern Cosmology and the Dark Matter Problem* (Cambridge Univ. Press, Cambridge, UK, 1993).

J. Silk, *A Short History of the Universe* (Scientific American Library, a Division of HPHLP, New York, 1997).

P. Sokolsky, *Introduction to Ultrahigh Energy Cosmic Ray Physics* (Westview Press, A member of the Perseus Books Group, Boulder Colorado, USA, 2004).

T. Stanev, *High Energy Cosmic Rays* (Springer-Verlag Berlin Heidelberg New York, 2004)

T. X. Trinh, *The Secret Melody* (Oxford Univ. Press, New York, 1995).

J. N. Wilford, *Cosmic Dispatches -- The New York Times Reports on Astronomy and Cosmology* (W. W. Norton & Company, New York and London, 2002).

W. S. C. Williams, *Nuclear and Particle Physics* (Oxford Univ. Press, Oxford and New York, 1991).

S. Yoshida, *Ultra-High Energy Particle Astrophysics* (Nova Science Publishers, Inc., New York N.Y., 2003).

GLOSSARY

(This glossary is also applicable to the author's December 2003 book, *How Dark Matter Created Dark Energy And The Sun.*)

Accretion: An infall of matter on an object.

AGASA: Akeno Giant Air Shower Array.

Alpha Particle: The nucleus of a helium atom.

Andromeda Galaxy: Twin galaxy of the Milky Way. Together, the two galaxies comprise most of the mass of the Local Group. Andromeda is also known as M31.

ASCA: Advanced Satellite for Cosmology and Astrophysics, which was on an X ray mission 1993-2001.

Astronomical Unit (A.U.): The average distance from the Sun to the Earth, equal to 149,598,000 kilometers.

Astrophysics: The study of the composition and other physical properties of celestial objects.

Astrophysical Emergence: See emergence and emergent evolution.

Astrophysical Dynamo Effect: Relativistic protons orbiting galaxies will create magnetic fields through the astrophysical dynamo effect under which the relativistic protons moving in Larmor orbits create magnetic fields.

These same magnetic fields in turn determine the proton paths, eventually reaching a steady-state solution for the magnetic fields and the proton paths after an emergent evolution period involving millions to billions of years.

"Attract"/"Attraction": A new term (in quotes) that refers to the movement in space of a UHE relativistic proton from one magnetic field strength to a higher magnetic field strength, resulting in the slowing down and dwelling in the region of the higher magnetic field.

Baryon/Baryonic: An elementary particle that is subject to the strong nuclear interaction. The proton and neutron are baryons.

Big Bang: The cosmological theory that holds that all the matter and energy in the Universe was concentrated in an immensely hot and dense point, which exploded 13.7 billion years ago.

Black Hole: An object that exerts such enormous gravitational force that nothing, not even light or other forms of electromagnetic radiation, can escape from it.

Bottom-Up Theory: The theory that small galaxies form first and larger galaxies are formed through mergers of the small galaxies.

CERN: Center for European Nuclear Research.

Chandra X-ray Observatory: Part of NASA's fleet of "Great Observatories" along with the Hubble Space Telescope and the Spitzer Space Telescope. Chandra allows scientists to obtain unprecedented X-ray images of exotic environments.

Closed Universe: A universe in which the density of matter is greater than the critical density and that will thus collapse onto itself in the future.

COBE: Cosmic Background Explorer.

Cold Dark Matter (CDM): Non-baryonic matter consisting of elementary particles of relatively high mass that are moving relatively slowly. (The term "cold" indicates a low temperature and thus a small energy of motion.)

Collective Self-organization: A term very similar in meaning to emergence and emergent evolution.

Collision Cross Section: A measure of the probability that an encounter between particles will result in the occurrence of a particular atomic or nuclear reaction.

Coma Cluster: A galaxy cluster that contains about 1,000 galaxies. The gravitational effects of dark matter were discovered in this galaxy cluster.

Comet: A body of ice and dust, with a nucleus of typically about 10 kilometers in diameter. It is visible only when it travels close enough to the Sun to reflect light.

Cosmic DM Mysteries/Cosmic Constituents: Dark matter mysteries or unexplained phenomena regarding celestial bodies or cosmic matter such as their shape, mass distribution, particle abundance ratios, dimensions, density, location, maturity, acceleration, velocity, linear momentum, angular momentum, particle energies, star rotation curves, hydrogen fusion reactions, particle energy distributions, particle transformations, star ignition, and star formation rates.

Cosmic Microwave Background (CMB) Or Cosmic Background Radiation (CBR): The microwave radiation that bathes the entire Universe and that dates from the epoch when the Universe was just 300,000 years old.

Cosmic-ray Cosmology: A new term to describe a cosmology recently developed by J. Drexler, based upon UHE protons and cosmic ray protons.

Cosmic Rays: Particles (mostly protons and electrons) that have been accelerated somewhere in the Universe to very high energies.

Cosmology: The study of the Universe as a whole, and of its structure and evolution.

Cosmos: An orderly, harmonious, and systematic Universe.

Coulomb Force: The force between two coulomb charges or electrically charged particles.

Dark Energy (as defined in the past): A hypothetical form of energy that permeates all space and has negative pressure resulting in a repulsive gravitational force. The accelerating expansion of the Universe has been attributed to dark energy.

Dark Galaxy: A galaxy with a total light output or luminous level from stars below an established minimum threshold level. Such galaxies represent extreme cases of low surface brightness (LSB) galaxies.

Dark Matter (as defined in the past): Matter that is detected only by its gravitational pull on visible matter. The composition has been unknown; it might consist of very low mass stars or supermassive black holes, but Big Bang nucleosynthesis calculations limit the amount of such baryonic matter to a small fraction of the critical mass density. If the mass density is critical, as predicted by the simplest versions of inflation, then the bulk of the dark matter must be a gas of weakly interacting non-baryonic particles, sometimes called WIMPs (Weakly Interacting Massive Particles).

"Death Spiral": A new term (in quotes) that refers to the spiral path of a UHE relativistic proton moving from a lower magnetic field to a considerably higher magnetic field wherein the radius of curvature of the UHE proton's spiral path is greatly reduced, leading to a significant increase in the synchrotron radiation losses and rapid decrease of the proton's kinetic energy.

Deuterium: A chemical element whose nucleus consists of a proton and a neutron, created mainly in the first three minutes of the Universe's history.

Doppler Effect: The variation in the energy and color of light caused by the motion of a source of light relative to an observer. If the source is receding, the energy decreases and the light is shifted toward the red. If the source is approaching, the energy increases and the light is shifted toward the blue.

Doppler Shift: The shift in the received frequency and wavelength of an electromagnetic wave that occurs when either the source or the observer is in motion. Approach causes a shift toward shorter wavelengths and higher frequencies called a blue shift. Recession has the opposite effect, called a red shift. The expansion of the Universe causes ancient electromagnetic wave emissions to exhibit a doppler red shift.

Dwarf Galaxy: A galaxy with a small size and mass. The average diameter is about 15,000 light-years; that is, about one-sixth of that of a normal galaxy. Masses range from 100 million to 1 billion solar masses, about 1,000 to 10,000 times less than the mass of an ordinary galaxy. Dwarf galaxies may be spheroidal or irregular, but dwarf spiral galaxies have not been observed.

Dwarf Spheroidal Galaxy (dSph): A dwarf galaxy that is spheroidal in shape, has an old stellar population, and lies close to a large host galaxy as a satellite.

Dwarf Irregular Galaxy (dIrr): A dwarf galaxy that is irregular in shape, has a young stellar population, and is a satellite of a large host galaxy, but lies further away from its host galaxy than would a dwarf spheroidal galaxy.

Electromagnetic Wave: A pattern of electric and magnetic fields that moves through space. Depending on the wavelength, an electromagnetic wave can be a radio wave, a microwave, an infrared wave, a wave of visible light, an ultraviolet wave, a beam of X rays, or a beam of gamma rays.

Electron: The lightest of the subatomic particles with electrical charge. The electron has a mass of 9×10^{-28} kilograms and is negatively charged.

Electron Volt (eV): The energy released when a single electron passes through a one-volt battery.

Elliptical Galaxy: A galaxy observed as an oval-shaped system generally composed of old stars, a large black hole, and containing little or no gas and dust.

Emergence, Emergent Evolution: A theory that new characteristics and qualities appear in the evolutionary process at more complex organizational levels (than that of the pre-existent entities such as a molecule, a cell, or a particle) and which cannot be predicted solely by studying less complex levels of organization but which are determined by a rearrangement of pre-existent entities.

Extragalactic, Intergalactic: The regions of the Universe outside of any galaxy.

Field Galaxies: The variety of galaxy types typically found in galaxy surveys.

Galactic Disk: A flattened aggregation of stars, gas, and dust in a spiral galaxy. The average disk is some 90,000 light-years in diameter and 300 light-years thick. In the Milky Way, the stars complete one turn around the galactic center every 250 million years, at a velocity of 230 kilometers per second.

Galactic Halo: A spherical region around a spiral galaxy populated by old stars and globular clusters. Observations suggest that it is surrounded by a dark matter halo some 10 to 20 times larger than the galaxy and more massive.

Galaxy: A system of stars (10 million in a dwarf galaxy, 100-200 billion in an average galaxy like the Milky Way, 10 trillion in a giant galaxy) held together by gravity.

Galaxy Cluster: A dense grouping of several thousand galaxies bound by gravity, with an average diameter of some 60 million light-years, and an average mass of a few million billion solar masses.

Gamma Ray: An electromagnetic wave with a wavelength in the range of 10^{-13} to 10^{-10} meters, corresponding to photons with energy in the range of 10^4 to 10^7 electron volts. Their energies are higher than X rays.

Gauss: A measure of the strength of a magnetic field.

General Relativity: A gravitational theory proposed by Albert Einstein in 1915, which is more accurate than that of Newton. The two theories differ mainly in situations where gravitational fields are very intense, such as around a pulsar or black hole. General relativity constitutes the theoretical support of the Big Bang theory.

GeV: G stands for Giga, or 10^9. Thus, GeV is one billion electron volts.

Gravitational Field: A field of force surrounding a body of finite mass. The field of force is defined as the force that would be experienced by a standard mass positioned at each point in the field.

Gravitational Tidal Force: The tidal force responsible for attraction between all matter. The weakest of the four forces, gravitational force possesses the longest range.

GRB: A gamma ray burst.

Great Attractor: A large grouping of galaxies with a total mass of 100 million billion solar masses, gravitationally attracting the Local Supercluster, which is moving toward it.

Great Wall: A sheet of galaxies which stretches more than 500 million light-years across the sky.

Group of Galaxies: A collection of about 20 galaxies held together by gravity, some six million light-years across and averaging between one and 10 trillion solar masses.

Gyr: Gigayear, or one billion years.

GZK Cosmic-Ray Cutoff: A theory limiting proton energies. According to the currently questioned 1966 Greisen-Zatsepin-Kuzmin (GZK) cutoff theory, protons with energies greater than 6×10^{19} eV would interact with the cosmic microwave background radiation and lose energy through radiation and thus would not travel more than 50 Mpc, or about 160 million light-years. In 1998, Coleman and Glashow wrote a paper entitled, "Evading the GZK Cosmic-Ray Cutoff," which showed that for very high energy cosmic rays, the GZK cutoff would not apply.

Halo: The region around a galaxy that contains dark matter and some halo stars.

Heavy Elements: All chemical elements with nuclei heavier than helium. Also known as "metals," these heavy elements are built up by nuclear fusion in the interiors of stars and supernovae.

Helium: A chemical element with a nucleus of two protons and two neutrons (helium-4). A second, far-less-abundant isotope has two protons and one neutron (helium-3).

Hubble Law: The law discovered in 1929 by the American astronomer Edwin Hubble, which states that the distance of galaxies varies in proportion to their red shift and, thus, because of the Doppler effect, to their velocity of recession. The law gave birth to the idea of an expanding universe.

Hydrogen: The lightest of all chemical elements, consisting of one proton and one electron. Hydrogen makes up 75% of the mass of the Universe.

Isotropy/Isotropic: The property of the Universe to be similar in every direction.

Kpc: The abbreviation for a kilo parsec where a parsec equals 3.26 light-years.

Large Magellanic Cloud (LMC): The larger of two irregularly shaped galaxies closest to the Milky Way located in the far southern sky and visible to the unaided eye.

Larmor Radius (for a proton): A proton crossing an orthogonal/magnetic field and entering into a spiral path. The radius of a cycle of that spiral path is called the proton Larmor Radius for that cycle.

$$\text{Proton Larmor Radius} = 110 \text{ Kpc} \times \frac{10^{-8} \text{ gauss}}{B} \times \frac{E}{10^{18} \text{ eV}}$$

Light-year: The distance traveled by light (which moves at a velocity of 300,000 kilometers per second) in one year and equal to 9,460 billion kilometers.

Local Group: A grouping of galaxies extending over a region of space of about 10 million light-years, of which the Milky Way and Andromeda are the principal and most massive members (one trillion solar masses each). It also includes dwarf galaxies.

Low-Surface Brightness (LSB) Galaxy: A diffuse galaxy with a surface brightness that is one magnitude lower than the ambient night sky.

Magnetic Bulge: A significant rise in the orthogonal magnetic field experienced by a relativistic proton.

Magnetic Field: A field of force in space, created by a magnet or by an electric current, that guides the trajectories of electrically charged particles by exerting an electromagnetic force.

Mass: The measure of the inertia of an object, determined by observing the acceleration when a known force is applied. An object with mass creates a gravitational field, which is defined in this glossary. When a proton travels at relativistic velocities, it has a relativistic mass equal to its energy divided by the square of the speed of light.

Maxwell's Equations: A set of differential equations describing space and time dependence of the electromagnetic field and forming the basis for classical electrodynamics.

Microwave: An electromagnetic wave with a wavelength of between one millimeter and 30 centimeters.

Milky Way: The galaxy to which our solar system belongs, whose central regions appear as a band of light or "milky way" that we can see from Earth in clear night skies.

Missing Mass: An outmoded name for the dark matter of the Universe.

MNRAS: Monthly Notes of the Royal Astronomical Society.

m_0: The symbol m_0 representing the mass of a proton when it is not moving (the rest mass).

Momentum: The linear momentum of an object, equaling the product of its mass and velocity. If no external forces are acting on a group of mass objects, the Law of Conservation of Linear Momentum requires that the total linear momentum of the mass objects in the group remains unchanged.

Muon (contraction of the earlier mu-meson; taken as a symbol for meson, and used to distinguish it from the short-lived pi-meson): An unstable elementary particle that belongs to the lepton family, that is common in the cosmic radiation near the Earth's surface, that has a mass about 207 times the mass of the electron, and that exists in negative and positive forms.

Muonic Ion: Two nuclei of atoms in close proximity, usually one of them being a proton and the other being a deuteron, a helium nucleus, or another proton, being orbited very closely by a single negative muon weighing 207 times as much as an electron at rest. Muonic ions are best known for catalyzing low temperature nuclear fusion reactions.

Neutralino: A theoretical non-baryonic particle, which is an amalgam of the superpartners of the photon (which transmits the electromagnetic force), the Z boson (which transmits the so-called weak nuclear force), and perhaps other particle types. Although the neutralino is heavy by normal standards (at least 35 times the mass of a proton), it is generally thought to be the lightest supersymmetric particle.

Neutron: A subatomic particle with no electric charge, one of the two basic constituents of an atomic nucleus.

Nucleosythesis: The production of a chemical element from hydrogen nuclei or protons, as in stellar evolution.

Ockham's Razor Logic (also called Occam's Razor): A scientific and philosophic rule that the favored explanation for an unknown phenomenon is the simplest of the competing theories. It should be preferred to the more complex, or that explanations of unknown phenomena be sought first in terms of known quantities rather than through assumptions.

Open Universe: A Universe in which the density of matter is less than the critical density and which will thus expand forever.

Oort Cloud: A region in the outer limits of the solar system where billions to trillions of comets reside.

Orthogonal: Intersecting or lying at right angles.

Parsec: An astronomical unit of distance equal to 3.26 light-years or approximately 19 trillion miles.

Pion (contraction of pi-meson): A short-lived meson that is primarily responsible for the nuclear force and that exists as a positive or negative particle with mass 273.2 times the electron mass or a neutral particle with mass 264.2 times the electron mass.

Population I Stars: A younger generation of stars with ages from a few million years to about 10 billion years and with a relatively large fractional abundance (about 1% of mass) of elements heavier than helium. The Sun is in this category.

Primordial Ripples: The mass perturbations in the early Universe that may have evolved into galaxies.

Proto-galaxy: A cloud of gas and ions that is evolving into a galaxy.

Proton: A positively charged particle composed of three quarks that, together with the neutron, forms atomic nuclei. The proton is 1,836 times more massive than the electron.

Recombination: In the traditional theory, between 300,000 and 700,000 years after the Big Bang, the plasma of free electrons and hydrogen nuclei that condensed to form a neutral gas, in a process called recombination. The prefix "re" is not meaningful here, however, since according to the Big Bang theory, the electrons and protons (hydrogen nuclei) were combining for the first time ever.

Red Shift: A shift to longer wavelengths and lower frequencies, typically caused by the doppler effect in a receding object or caused by the expansion of the Universe.

Reductionism: A procedure or theory that reduces or attempts to reduce complex data or phenomena to simple elements or terms. It is an inward-looking approach that in physics usually means a search for subatomic particles in an attempt to understand or explain some unusual phenomenon.

Relationism: An analytical procedure, method, concept, or theory, developed by Jerome Drexler, that attempts to identify dark matter by determining which cosmic phenomena (called Cosmic DM Mysteries) may be facilitated, expedited, influenced by, or have a special relationship with dark matter. This outward-looking cosmological concept is used to determine the nature and characteristics of dark matter's influence on and relationship with cosmic phenomena as a means of DM identification.

Rotation Curve of a Galaxy: A graph of the orbital velocities of stars or hydrogen as a function of their radial distances from the nucleus of the galaxy radially outward into the surrounding dark matter halo.

Schmidt Law: An empirical law, for isolated spiral galaxies, of the correlation between the star formation rate (SFR) and the overall average molecular hydrogen surface density.

SFR: Star formation rate.

Signature Characteristics (SigChar): An extensive list, for each dark matter candidate, of all the possible features or characteristics of the candidate that can be attributed to it by utilizing any and all laws and principles of physics and chemistry.

Solar System: The Sun and the objects in orbit around it, which include nine planets, nearly 60 known satellites of the planets, thousands of smaller objects called asteroids, and billions to trillions of comets.

Solar Wind: Astronomers estimate that about 20% of the Sun's mass has been lost due to the solar wind.

Spiral Galaxy: A flattened, disk-like system of stars and interstellar gas and dust with a spherical collection of stars, known as the bulge, at its center. Bright, young stars outline spiral arms in the plane of the disk.

(The) Standard Model: Name given to the current theory of fundamental particles and how they interact.

Star: A sphere of gas consisting of 98% hydrogen and helium and 2% heavy elements in equilibrium under the action of two opposing forces -- the compressive gravity and the outward radiation pressure from the nuclear fusion reactions in its core. The Sun has a mass of 2×10^{30} kilograms, and masses of stars range between 0.1 and 100 solar masses.

Starburst Galaxy: A galaxy experiencing a period of intense star forming activity. They are usually associated with the merging or interaction of two galaxies. This activity may last for 10 million years or more. During a starburst, stars can form at tens, even hundreds, of times greater rates than the star formation rate in normal spiral galaxies.

Supercluster: The aggregation of tens of thousands of galaxies held together by gravity and gathered into groups and clusters. Superclusters have the shape of flattened pancakes with an average diameter of 90 million light-years and masses of 10,000 trillion (10^{16}) solar masses.

Supernova/Supernovae: An exploding star, visible for weeks or months, even at enormous distances, because of the tremendous amounts of energy that the star produces. Supernovae typically arise when massive stars exhaust all means of producing energy from nuclear fusion. In these stars, the collapse of the star's core results in the explosion of the star's outer layers. Another type of supernova arises when hydrogen-rich matter from a companion star accumulates on the surface of a white dwarf and then undergoes nuclear fusion. This second type, known as a Type 1a supernova, generates light at a well-known standard level and thus can be used to measure the rate of expansion of the Universe.

Synchrotron Radiation: Electromagnetic radiation that is emitted by charged particles moving at relativistic speeds in circular orbits in a magnetic field. The rate of emission is inversely proportional to the product of the radius of curvature of the orbit and the fourth power of the mass of the particles. For this reason, synchrotron radiation is not a problem in the design of proton synchrotrons, but it is significant in electron synchrotrons.

TeV: T stands for Tera, or 10^{12}. Thus, TeV is one trillion electron volts.

Tokamak: Tokamak Fusion Test Reactor (for fusion of hydrogen isotopes), which by 1997 achieved a world record plasma temperature of 510 million degrees centigrade -- the highest ever produced in a laboratory and well beyond the 100 million degrees required for commercial hydrogen fusion.

Top-Down Theory: The theory that long, large, dark matter filaments form galaxy clusters where the DM filaments intersect/collide and then galaxies form from the remnants of these collisions.

UHE (Ultra-High Energy) Proton: A proton traveling near the speed of light with an energy of at least 10^{18} eV.

UHECR: Ultra-high energy cosmic ray proton traveling near the speed of light with an energy of at least 10^{18} eV.

Ultraviolet (UV): Ultraviolet light.

Virgo Supercluster: A huge, flattened supercluster that contains the Local Group of galaxies. The Local Group, containing the Milky Way, lies at the edge of the supercluster, while the Virgo Cluster of galaxies is at its center.

White Dwarf Star: A small, dense star with a diameter of about 10,000 kilometers (about the size of Earth) created when a star of less than 1.4 solar masses exhausts the nuclear fuel and collapses under its own gravity. This type of star participates in a Type 1a supernova.

WIMP (Weakly Interacting Massive Particle): The name for a non-baryonic theoretical dark matter candidate that is presumed to have a mass much greater than that of a proton. A neutralino is one form of WIMP.

WMAP (Wilkinson Microwave Anisotropy Probe): A NASA Explorer mission measuring the temperature of the cosmic background radiation over the full sky. This map of the remnant heat of the Big Bang provides data about the origin of the Universe.

X Rays: Electromagnetic radiation with greater frequencies and smaller wavelengths than those of ultraviolet radiation and lower frequencies and longer wavelengths than those of gamma ray radiation.

INDEX

Accelerating expansion, x, xi, 18, 21, 23, 31, 36, 37, 38, 71, 72, 73, 91, 92, 93, 107, 110, 114, 119, 120, 152, 155, 212, 215, 244, 245, 246, 247, 272

Accretion, 24, 86, 89, 113, 127, 180, 181, 193, 209, 216, **269**

Aloisio, R., 239, 263

Alvarez, Luis W., 100, 148

Ankle, x, xiii, 18, 21, 23, 39, 69, 107, 110, 149, 150, 151, 165, 215,

Anchordoqui, Luis A., 174, 175, 176, 262

Andromeda galaxy, front cover, 24, 117, 177, 181, 216, 221, 228, 244, **269**, 279

Astrophysical dynamo, 40, 45, 53, 92, 114, 123, 131, 259, 267, **269**

Astrophysical emergence, xiv, 41, 42, 118, 129, 130, 155, 157, 158, 160, 162, 212, 218, 229, 226, **269**

Bahcall, John N., 263, 265

Baryon/baryonic, xv, 25, 40, 41, 44, 80, 112, 126, 142, 180, 195, 196, 197, 198, 199, 212, 220, 235, 236, 237, **270**, 273

Berezinsky, V. S., 239, 263

Bishop, A. S., 261, 265

Blumenthal, George, 233, 259

Bradner, H., 110, 148

Bundy, Kevin, 208, 262

Carnegie Observatories, 79, 80, 127, 128, 261

CERN, 243, **270**

Chandra x-ray images, 77, 125, **270**

Clay, R., 261, 265

CMB (Cosmic Microwave Background), 27, 35, 112, 155, 199, 203, 235, 238, **272**

COBE (Cosmic Background Explorer), 35, 235, **271**

Coleman, Sidney, 239, 263, 277

Collective self-organization, 42, 114, 156, 158, 159, 160, 161, **271**

Coma cluster, v, **271**

Coulomb's law, 112

Coulomb force, 26, 105, 113, 118, 133, 189, 190, **272**

Cronin, James W., 115, 239, 261

Croswell, Ken, 88, 265

Dark energy, vi, xi, xvi, 38, 72, 73, 93, 107, 119, 120, 224, 233, 241, 244, 259, 265, 266, 269, **272**

Dark galaxy, 85, 88, 89, 135, 138, **272**

Dark matter relationism, vii, 20, 33, 109, **186**, 190, 194, 199, 202, 206, 213, 214, **282**, 294

Dawson, B., 261, 265

Deuterium, 98, 100, 104, 146, 148, 235, **273**

DOE (U.S. Department of Energy), vi

Downsizing, 209

Dwarf galaxies, x, xiii, 18, 21, 23, 37, 39, 65, 85, 86, 87, 88, 95, 106, 107, 110, 135, 137, 138, 143, 152, 170, 187, 189, 210, 211, 212, 215, 223, 227, 263, **274**, 275, 278

Dwarf irregular galaxies (dIrr), 229, **274**

Dwarf spheroidal galaxies (dSph), 221, 229, 230, 231, **274**

Ellis, Richard S., 208

Emergence, emergent evolution, xiv, 40, 41, 42, 45, 114, 118, 123, 129, 130, 155, 156, 157, 158, 159, 160, 161, 162, 212, 218, 220, 226, 270, 271, **275**

ESO (European Southern Observatory), 95, 103

Extragalactic magnetic fields, 37, 39, 43, 45, 47, 71, 81, 115, 117, 129, 135, 155, 175, 184, 185, 186, 227, 250, 253, 254, 255

Faber, Sandra, 233, 259

Farrar, Glennys, 164, 165, 174, 176, 262

Fornax cluster/Fornax galaxy, 77, 125, 229, 231

Gilmore, Gerry, 211, 221, 222, 223, 224, 225, 226, 227, 228, 231

Glashow, Sheldon L., 239, 263, 277

Griest, Kim, 233

GZK, 238, 239, 240, **277**

Hillas, A.M., 259

Hubble Law, 72, 119, **277**

Hydrogen fusion, xiii, 17, 18, 19, 21, 23, 39, 63, 88, 97, 98, 99, 100, 101, 102, 104, 105, 110, 135, 143, 145, 146, 148, 170, 172, 178, 180, 182, 194, 203, 204, 206, 215, 217, 218, 272, 285

I Zwicky 18/I Zw 18, 89, 138, 228

Keel, William C., 168

Knees, x, xiii, 18, 21, 23, 39, 69, 107, 110, 149, 150, 151, 165, 215, 243, 255

Kravtsov, Andrey V., 262

Larmor Radius, ix, 38, 41, 47, 48, 49, 51, 71, 81, 92, 106, 112, 113, 117, 129, 149, 155, 161, 175, 179, 184, 185, 198, 228, 250, 251, 253, 254, 255, **278**

Linear momentum/momentum, 40, 72, 73, 114, 119, 245, 246, 271, **280**

LMC (Large Magellanic Cloud), 53, **278**

Local Group, xvi, 22, 48, 49, 50, 70, 75, 110, 149, 150, 151, 189, 221, 231, 244, 269, **278**, 285

LSB (Low Surface Brightness) galaxy, x, xi, xiii, 18, 21, 23, 37, 39, 65, 66, 85, 85, 87, 88, 106, 107, 110, 135, 137, 138, 152, 155, 187, 190, 210, 211, 212, 215, 227, 272, **279**

Magnetic bulge, x, 59, 60, 106, 256, 257, 258, **279**

McGaugh, Stacy S., 195, 196, 197, 198, 199, 262

Milky Way, viii, ix, xv, 18, 43, 45, 47, 48, 49, 50, 68, 69, 70, 75, 86, 87, 89, 106, 110, 111, 113, 121, 133, 135, 136, 149, 150, 151, 183, 184, 185, 189, 210, 221, 228, 229, 244, 249, 250, 253, 254, 255, 258, 269, 275, 278, **279**, 285

Minchin, Robert, 223, 260

Muon, xiii, 63, 79, 86, 88, 97, 98, 99, 100, 101, 102, 103, 104, 105, 112, 122, 141, 143, 145, 146, 147, 148, 152, 155, 162, 170, 172, 180, 193, 194, 199, 203, 204, 205, 206, 212, 217, **280**

Muonic ion, 88, 99, 101, 102, 147, 180, 194, 204, 205, 206, 217, **280**

NASA, vi, 38, 77, 125, 218, 219, 261, 270, 286

Neutralino, xiv, 28, 29, 35, 153, 154, 214, **280**, 281, 286,

Noeske, Kai G., 208

Non-baryonic, 27, 80, 126, 128, 131, 271, 273, 280, 286

NSF (National Science Foundation), vi

Nucleosythesis, 235, 236, **281**

Ockham's razor/Occam's razor, 29, 30, 213, 220, **281**

Ostriker, Jeremiah P., vii, 263, 265

Pen, Ue-Li, 188, 262

Pierpaoli, Elena, 164, 165, 174, 176, 262

Pion, 98, 122, 146, 162, 193, 203, **281**

Primordial star formation, 95, 96, 99, 143, 144, 147

Proto-galaxy/galaxies, 79, 113, 127, 193, 209, 257, **282**

Qin, Bo, 188, 262

Reductionism, 20, **282**

Rees, Sir Martin, 115, 233, 259, 261

Relationism, vii, xiii, xvi, 17, 20, 21, 22, 33, 36, 109, 152, 185, 186, 190, 194, 199, 203, 206, 212, 213, 214, 220, **282**

Rotation curves of galaxies, x, xiii, 19, 31, 35, 37, 39, 65, **66**, 68, 106, 137, 139, 152, 212, 272, **283**

Rubin, Vera, v, 31, 66, 67, 68, 139, 140, 259, 267

Schmidt Law, xv, 23, 87, 182, **201**, **202**, 203, 205, 206, 211, 212, 215, **283**

Self-interacting/self-interaction, 23, 129, 130, 155, 187, 188, 189, 190, 197, 215, 225, 226, 227

SFR (Star Formation Rate), xi, xiii, xiv, xv, xvi, 23, 85, 86, 87, 88, 107, 110, 113, 114, 138, 155, 167, 168, 171, 172, 182, 201, 207, 208, 209, 210, 211, 212, 215, **283**

Silk, Joseph, 188, 262, 267

Sokolsky, Pierre, 91, 260, 267

Special Theory of Relativity, v, 41

Spergel, David N., 188, 189, 262

Spiral galaxy, xiv, 24, 47, 66, 67, 68, 83, 89, 100. 117, 119, 131, 135, 139, 140, 144, 167, 168, 169, 170, 171, 172, 173, 175, 176, 177, 178, 179, 183. 187, 190, 196, 203, 204, 216, 249, 251, 254, 275, **283**

Stanford SLAC, 103

Starburst galaxy, xiv, 24, 56, 95, 147, 99, 167, 168, 170, 171, 172, 173, 174, 176, 202, 211, 212, 216, 262, **284**

Starless galaxy, xiii, 39, 86, 135

Steinhardt, Paul J., vii, 188, 189, 262

Strongly interacting, 111, 114

Synchrotron radiation, ix, x. 41, 43, 46, 50, 51, 57, 59, 61, 70, 71, 75, 84, 85, 106, 107, 113, 114, 119, 121, 122, 124, 134, 136, 155, 162, 179, 193, 203, 228, 229, 230, 245, 246, 247, 249, 250, 251, 252, 256. 257, 273, **285**

Teller, Edward, 100, 148

Tokamak, 100, **285**

Top-down theory of galaxy formation, 77, 111, **113**, 126, 127, 128, 199, 209, 219, **285**

Torres, Diego F., 174, 175, 176, 262

Trinh, Thuan Xuan, 268

UHE proton, 49, 59, 60, 70, 239, 245, 246, 247, 248, 249, 250, 251, 252, 255, 256, 257, 258, 272, 273

UHECR, xiv, 24, 163, 164, 166, 170, 173, 174, 175, 176, 212, 216, **285**

Van Zee, Liese, 87

Virgo supercluster, 22, 49, 50, 70, 110, 149, 150, 151, 163. 165, 166, **285**

Watson, A. A., 115, 233, 259, 261

Wilkinson, Mark I., 35, 211, 222, 227, 231, 263

WIMP, v, vi, xiv, 28, 29, 35, 80, 81, 126, 130, 132, 153, 154, 160, 162, 214, 218, 219, 273, **286**

WMAP, 35, 223, 225, **286**

Yoshida, S., 260, 268

Zwicky, Fritz, v, 31, 259

Copyright © 2006 Jerome Drexler, Los Altos Hills, California.
All rights reserved.

Comprehending And Decoding The Cosmos

DISCOVERING SOLUTIONS TO OVER A DOZEN COSMIC MYSTERIES BY UTILIZING DARK MATTER RELATIONISM, COSMOLOGY, AND ASTROPHYSICS

There are many mysteries involving cosmic phenomena. Jerome Drexler used 14 of these and his analytical concept of *dark matter(DM) relationism* to discover a promising candidate for *dark matter*, the source of ultra-high energy cosmic rays, and theories for star formation, starburst galaxies, and the *emergence* of DM halos.

To test the validity of his discoveries, Drexler used another 11 unexplained cosmic phenomena discovered by astronomers primarily during 2005. Utilizing his same promising *dark matter* candidate, Drexler was able to explain in a plausible manner all 11 of these recently discovered cosmic mysteries.

Drexler's research has led not only to an identification of *dark matter* and to plausible explanations for the 25 cosmic phenomena, but also to a deeper understanding of many aspects of the cosmos, leading to a partial decoding of the cosmos.

About the Author

Jerome Drexler, a former Research Professor in the Department of Physics at the New Jersey Institute of Technology and author of two astrophysics/cosmology books, began his career as a Member of the Technical Staff of Bell Laboratories. He has been granted 76 U.S. patents involving megawatt microwaves, laser recording, the LaserCard® optical memory card, and nanotechnology. He founded LaserCard Corporation (Nasdaq: LCRD) and has received honorary degrees from the Technion-Israel Institute of Technology, New Jersey Institute of Technology (D.Sc.), and Upsala College (D.Sc.) and the "Inventor of the Year Award" in 1990 for California's Silicon Valley. He resides in Los Altos Hills, California.

Copyright © 2006 Jerome Drexler

www.ingramcontent.com/pod-product-compliance
Lightning Source LLC
Chambersburg PA
CBHW020731180526
45163CB00001B/195